MW00577587

PRACTICAL FLOOR MALTING

By
HUGH LANCASTER

London
"The Brewing Trade Review"
13 Little Trinity Lane, E.C.
1908

First edition originally published 1908

Printed in the United States of America.

ISBN 978-0-9910436-5-1

whitemulepress.com
cheers@whitemulepress.com

PO Box 577
Hayward, CA 94541

CONTENTS

I. Site And Construction 1
II. Barley Storage 20
III. Machinery 32
IV. Barley Drying 45
V. Steeping And Construction Of Floors 54
VI. Flooring 64
VII. Kiln Construction And Drying 85
VIII. Loading, Drying, And Curing 99
IX. Barley And Malt 114
X. Malting Loss 127
XI. Malting Cost 134
XII. Barleys 146
Bibliography 170
Index 172

PREFACE

As its title implies, "Practical Floor Malting" is intended to appeal to the practical brewer and maltster rather than to the theorist, and with this object in view theory has been avoided as much as possible throughout.

My hearty acknowledgments are due to Dr. E. R. Moritz, to Mr. J. D. Sandars, and to Mr. Thos. Haynes; also to Mr. E. S. Beaven, to whose work on Kilning and Barley special reference has been made, and to Mr. W. H. Phelp, for the care bestowed by him on the photographs of barley.

HUGH LANCASTER.

LONDON, *July* 13, 1908.

CHAPTER I

SITE AND CONSTRUCTION

Site of Maltings— Beyond the importance of its being situated as close as possible to the supply of raw material, little need be written upon the choice of a site for the erection of maltings. Sufficient is not yet known about the relative effects of hard and soft waters to make one or the other a necessity, and although low rates, cheap power, and cheap labour are all desirable, they must as a rule be somewhat insignificant factors where the position of the malthouse has to be determined by propinquity to the source of barley supply, or to the brewery or breweries for which malt is to be made. Where it is possible to secure a canal or river frontage, advantage should be taken of the comparative cheapness of water carriage, and if a large proportion of foreign barley is to be malted this is specially important; but it is also important from the fact that large quantities of good quality English malting barley are grown in counties on the coast, cargoes of which can be secured at low water rates in seasons when the barley grown near the maltings is not of satisfactory quality, and railway rates over long distances would make its purchase at a satisfactory price impossible.

Sharply rising ground or high buildings in close proximity to the kilns are likely to cause down-draught, and should therefore be avoided; and in setting out the ground plan care should be taken that the fronts of the kiln fires or furnaces should, if possible, face the quarter from which the prevailing wind blows, unless forced draught is to be used. A good draining soil is preferable to one that retains moisture, especially if the bottom working floor is sunk two or three feet below the ground level, as these floors generally get less fresh air than the others, and the combination of damp and want of ventilation in such floors may cause a tendency to mould. Perhaps, again, it is slightly preferable that the length of the floors should be from east to west rather than from north to south, as the drying winds are one of the greatest enemies to young pieces; and for the same reason a malthouse which stands alongside a river or canal has a slight advantage. Few things are more likely to encourage the spread of mould on the growing pieces than road dust, so that the windows of the growing floors should be kept away as much as possible from thoroughfares where the combination of dust and manure is likely to be a source of infection in dry windy weather.

In building brewery maltings it should not be forgotten that in the processes of screening both barley and malt a good deal of dust is formed, and that if the maltings are in close proximity to the brewery grave danger of wort infection may arise from such dust, containing as it does putrefactive germs rubbed from the husk of the barley and malt. No space need be wasted here by further alluding to the dangers of bacterial infection, but if it is necessary that the maltings should be close to the brewery the greatest precautions should be taken that all malt or barley dust should be sludged, burnt, or buried.

Effect of Type of Barley to be malted on Construction of House— The site having been determined, many points must be considered before plans are adopted, if the greatest possible economy is to be attained. There are several systems of malting on the floor system suitable to different kinds of barley and the production of different types of malt, and a house built for one of these systems will not necessarily work economically on another. At the present time most maltings not built within the last fifteen or twenty years are deficient in kiln area, while some more recently constructed houses have more kiln area than can be used owing to restricted flooring room. Some brewers, again, have a preference for slow grown malt which has been kept on the floors for thirteen or fourteen days, while others get better results from a flooring period of nine to eleven days, the latter system obviously requiring less floor area per quarter than the former. Barleys vary considerably in the flooring period they require, and with English this is chiefly a question of maturation in different seasons; but a house built for brewing Californian or Ben-Ghazi might well be allowed less floor space than one destined for Smyrnas, Ouchacs, or Brewing Chilians, which commonly take two or three days longer to modify than the first named varieties.

Economical Necessity of Working Maltings to fullest possible extent— Again, it should be remembered that malting expenses are to be classed under two headings, viz., those such as labour, water, power, interest on working capital, &c., which depend on the number of quarters malted, and, on the other hand, those such as rent, rates and taxes, and depreciation on buildings and plant, which are fixed and irrespective of the number of quarters of malt made up. Rates and taxes are dependent upon rent, and it must be noted that the cost of building large modern houses may vary from £80 to £120 per quarter on the steeping capacity at each wetting, that rent and depreciation on buildings should be charged at least 5 percent on the capital expended, and that the cost for rent and depreciation on buildings alone per quarter of malt made will be the total number of quarters of malt made each season divided into the year's rent. Taking all these matters into consideration, it at once becomes obvious that too much care cannot be taken to plan a building adapted for the production of the maximum amount of soundly-made malt at the minimum outlay.

Before leaving this subject it should be pointed out that houses malting thin foreign barleys which may be conveniently termed brewing barleys—such as Smyrna, Brewing Californian, Ouchac, &c. or a proportion of these have, in this respect, a great advantage over those whose make is restricted to English, or barleys such as French, Saale, Hungarian, Bohemian, Chevallier Chilian and Chevallier Californian. These latter may be conveniently classed as foreign Chevalliers, and can only be malted satisfactorily in average seasons between Sept. 30 and May 15. The brewing foreigns can generally be made up between Sept. 1 and June 15, so that in the case of a house capable of steeping 100 quarters every five days the maximum permissible season's make of English or Chevallier foreign would be 5000 quarters, while if a proportion of brewing foreigns were malted 6000 quarters could be malted. Now, if the rent of the-house were £500 per annum inclusive of depreciation on buildings, the cost of rent arid depreciation on buildings would be in the former case 2*s*. and in the latter 1*s*. 8*d*. per quarter.

Floor Space per Quarter of Malt made— The amount of floor space required for each quarter of malt to be made depends, as has been shown, on the quality of barley to be worked and the length of time it may be thought desirable to keep the grain on the floors and, of course, on the frequency with which barley is steeped. In many old-fashioned houses where emptying the cisterns is one of the hardest pieces of work, it is customary, especially in the south and south-east of England, to steep three times a fortnight, so that an emptying day should never fall on a Sunday; but with well-arranged conical cisterns emptying need take very little time and labour, so that Sunday emptying may be said to have lost its terror. Normal English barley may require a steeping period of from 48 to 65 hours, but the drier foreign sorts are often better for 72 or even up to 90 hours. If it be arranged to drain off the steep liquor overnight, and to leave the barley lying in the Cisterns for twelve hours between draining and emptying, improved drainage results, and in very cold weather the start of germination is slightly accelerated. This system of overnight draining, though not always permissible in the warmer months, has distinct advantages in winter; twelve hours should be allowed for it, so that with the time taken up in emptying, cleaning out the cisterns and getting barley in, four-day intervals become necessary, at any rate for houses which are to make up brewing foreign. Four-day steeping intervals, however, mean running the drying and curing processes somewhat fine, unless the kilns are built specially with that object in view, or forced draught or mechanical turners are employed, and it will probably be found better all round to arrange a new house to steep at five-day intervals, and proportion the steeps, floors, and kilns accordingly. If ten days be taken as the best average flooring period to fix, from the time of emptying the cisterns to the time of loading the kilns, this five-day system will leave only two pieces on each floor at any time, the older piece loading on the same morning as the cistern is emptied, giving the intermediate piece the benefit of nearly the whole floor on the sixth morning, when germination will in all probability be at its strongest. Under this system about 175 square feet of floor space will be required for each quarter of malt made. This figure has hitherto generally been given in square feet per quarter of barley steeped, but in considering houses adapted for either English or foreign barleys when the out-turn of malt from barley may vary between -2 and +10 percent, it would seem better to give figures for malt which is a constant factor than for barley which may absorb very varying amounts of water during the steeping process, and consequently vary enormously in bulk in the growing stage.

It may make matters clearer here if the case of a 200-quarter house is taken, this being the largest quantity which it is advisable to steep at a time in one building, and which may be represented by 180 to 205 quarters of barley. The floor space required on the five-day system will be about 8750 square feet, which may be divided into four, five, or six working floors, according to the height desired for the kilns and other circumstances; but it must be remembered that the width of the floors should be much less than the length in order to ensure ventilation and to give room for plenty of turning. If on an average two turns a day are required, the floors are kept out for ten days, and some six feet are gained by each turn, some 120 feet will be gained by turning; it is a pity to have to turn floors back—i.e., away from the kiln—unless absolutely necessary, and a good rule for the proportion of floors is two of length to one of breadth.

For convenient working the 200 quarters should be split up into three or four parts, each part having its own floor or floors, cistern and kiln, and, where possible, four floors corresponding with four cisterns and four kilns is a good arrangement; for if the building be designed on sound lines and the machinery well adapted for saving labour each maltster should be capable of looking after 25 quarters every five days, so that two men can be apportioned to each floor.

Kiln Space per Quarter of Malt made— The arrangement of the kilns invariably represents difficulties. They should be placed, if furnaces are used and furnaces are more economical of fuel than baskets so that the prevailing wind blows into the furnace, but they must also be as near as possible to the working floors and grouped as close together as possible for convenience in loading. Beyond this—

and the point is often overlooked—they should be arranged to save as much labour as possible in throwing the finished malt off them. Well-constructed kilns designed for five-day dryings should have areas sufficient to allow 21 square feet per quarter of malt, and such kilns will load to a depth of about twelve to thirteen inches of green malt. Some maltsters consider this depth too great for efficient, drying, and if it be found that the above thickness of green grain unduly retards the draught on the first day advantage may be taken of the system introduced by Mr. Thomas Haynes, jun., of Radcliffe-on-Trent, of loading in two halves, the first half being fairly dry after 24 hours, when the second half is loaded.

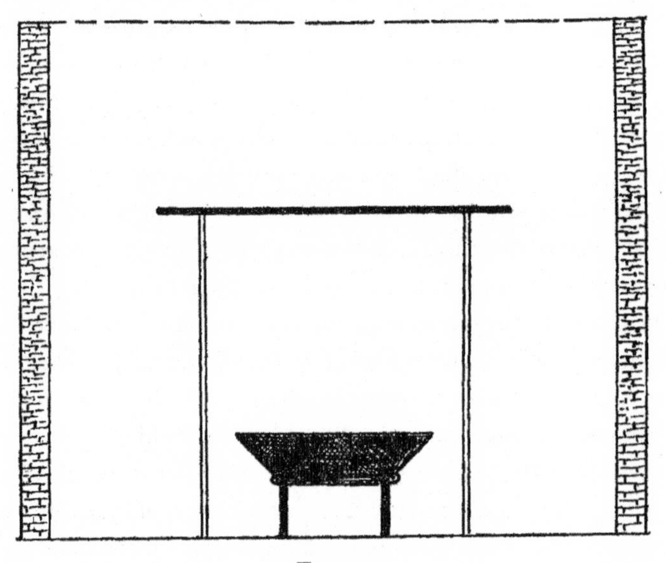

Fig. 1.

Malt kilns capable of dealing comfortably with 50 quarters every five days will require floor areas of about 1050 square feet, and although it is far best to have one fire only in a kiln of this size, and for that reason they should be square in shape rather than oblong, in order that the source of heat may be as central as possible, yet if it be found possible to get equal distribution with oblong, kilns, the labour of emptying them will be lightened to a very considerable extent.

Kiln Construction— It may be well here to give a brief description of the principles upon which malt kilns have been and are built, as the form the kiln takes is not without influence on the rest of the house, especially with regard to malt storage. The simplest form of kiln (Fig. 1) known in this country is a rectangular room some fourteen feet high from the ground floor to the drying floor, which is composed of perforated tiles; in the middle or in as central a position as possible of the ground floor is a cast-iron fire basket some two feet

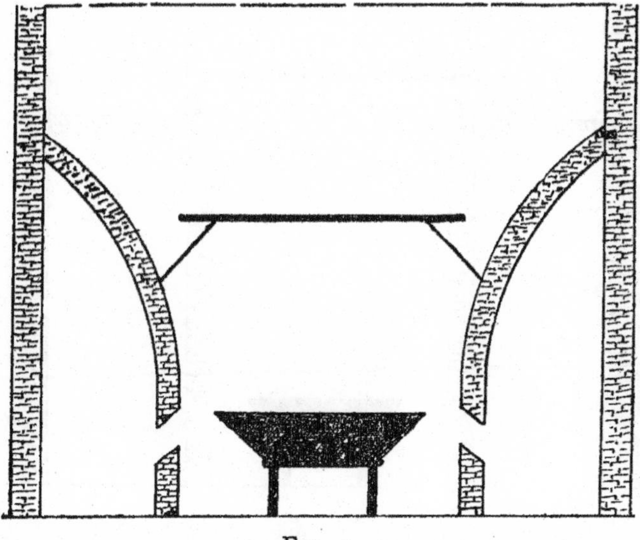

FIG. 2.

in height to hold the fire; six or seven feet above the fire, and supported by brick or iron columns, a sheet of cast iron, technically known as the "disperser," is laid horizontally over the fire protruding some distance beyond it on each side. It is not proposed to deal here with the structure above the drying floor, and in this somewhat primitive type of kiln there is nothing else below it. There is, or was within the last four years, at least one malt kiln of this type at work in England, and there may be more of them being used at the present time.

The commoner type of small kiln, however (Fig. 2), has a shaft built round the fire-basket, carried up and out above the fire to the outside walls in the form of half arches of brick, forming an inside chamber immediately below the drying floor known as the hot-air chamber.

In the larger and recently constructed kilns hardly any change has been made in principle, though for economy in construction the brick arch is sometimes replaced by a floor, parallel to the drying floor, and comparatively lightly built (Fig. 3), or by galvanized iron

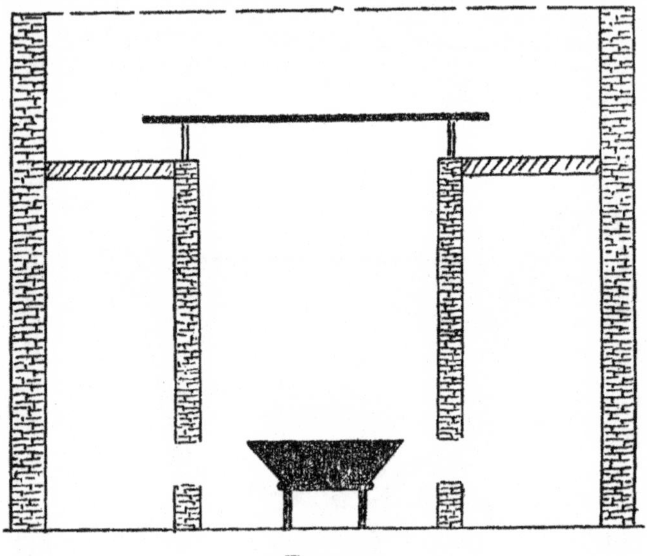

FIG. 3.

sheets put in at an obtuse angle with the top of the shaft (Fig. 4). In another type of old kiln (Fig. 5), equally common with the fire-basket type, the fire-basket is replaced by a bricked-in furnace, and in this construction (as in Fig. 5) the shaft is often simply a continuation of the furnace walls, and is consequently of much smaller area than the shaft of the other type of kiln. The small furnace kilns have their descendants in the modern shaft-kilns, though in many cases (Fig. 6) the sheet type of disperser, which in modern kilns is generally formed of tile or concrete, has given place to a dome; this is built upon the top of the shaft from which concentric arms or pipes, running from the dome to the outside walls, distribute the hot air over the whole area to be dried through spaces left in them for that purpose. Whereas in the modern fire basket type of kiln the shaft may take up as much as one-eighth of the kiln area, in the furnace type the shaft seldom takes more than one-fortieth to one-fiftieth of the kiln area, and this has led to the system of building malt stores below the hot-air chambers in many shaft kilns, the shaft walls being built double with an air space between to prevent undue radiation of heat.

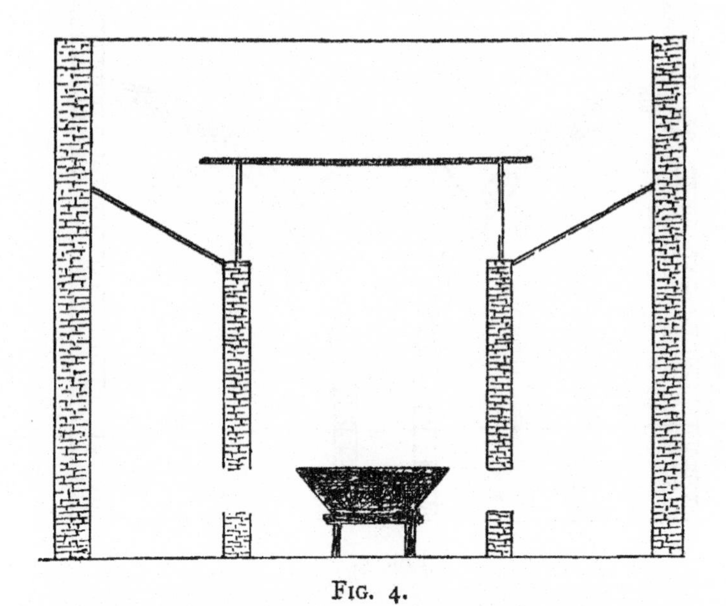

Fig. 4.

Malt Storage under Kilns— In high modern kilns, where the space between the ground floor and the drying floor may be anything up to 30 feet, malt stores, 8 or even 10 feet high and of the same superficial area as the kiln less the space taken up by the shaft, may easily be sandwiched in between the hot-air chamber and the stokehole, without unduly diminishing the height of the hot-air chamber. These malt stores may be capable, according to the height of the kiln, of holding seven to ten weeks' make of malt, and are placed in a part of the building which would otherwise be wasted room. A good deal of labour is also saved by this arrangement in the process of throwing the malt off the kilns, it being possible to drop the malt direct into the store through, say, four pipes connecting the kiln floor with the store through the hot-air chamber. The part of the drying floor taken up by such pipes is, of course, practically idle as regards drying, but as they need not be of more than one square foot each in area the space lost in kilns of anything beyond 900 square feet in

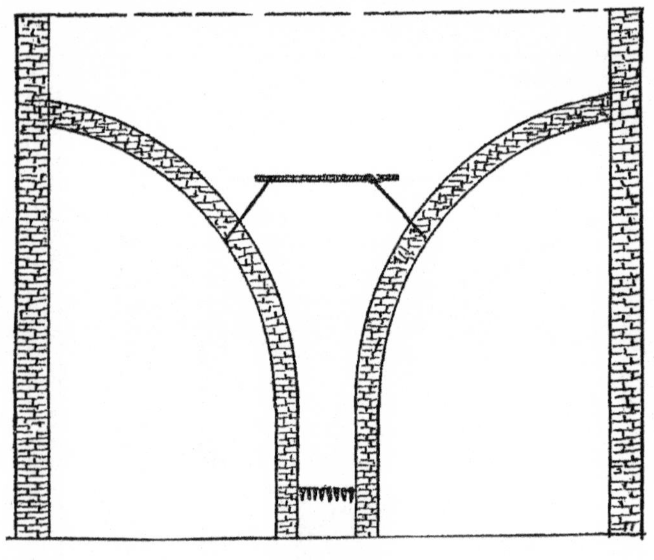

Fig. 5.

area will be hardly appreciable. It is, however, practically impossible to screen the malt as it is thrown off without very greatly retarding the throwing-off process, and malt and rootlets have consequently to be stored together and screened before being sent out, such screening and sacking having to be done by hand and entailing a good deal of time and labour.

There is much less inducement to use the kiln space as a malt store in the larger shaft-kilns, and in these it has become customary to build hoppers beneath the hot-air chamber either in one corner of the kiln or preferably, as regards labour-saving in throwing off, along the longest side of the kiln. By arranging these hoppers to command conveying machinery the kiln loads can be screened and stored in bin in the intervals between throwing off the successive kiln loads, without manual labour, and if hopper-bottomed malt bins are used commanding a movable automatic sacking-up and weighing apparatus, all that need be done once the malt has been thrown off the kiln will be to tie the sacks and run them on a sack barrow to the dray or truck.

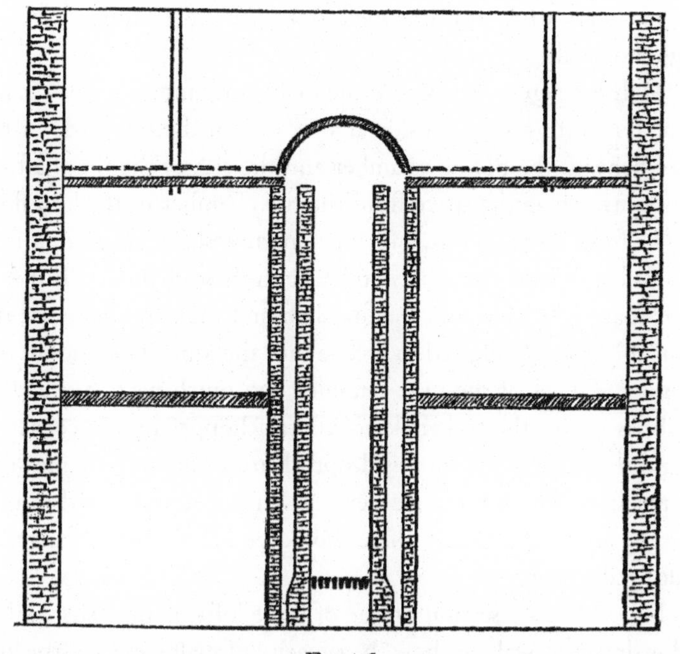

FIG. 6.

Before leaving the subject of kiln construction in its relation to the working of the malthouse, some mention should be made of double or, as they are sometimes called, dumping kilns. In these there are two floors, the bottom being preferably of tile and the top of wire; the former being intended for the curing and the latter for the drying of the malt. They are much used on the Continent and in parts of Scotland, but are not looked upon with favour in England owing to the difficulty of regulating the draught so as to maintain adequate velocity and a sufficiently low temperature on the top kiln, while there is a high temperature and practically no current of air passing through the bottom one. In order to attain this, a very careful arrangement of draught entering above the malt surface on the lower kiln is necessary. The merits and demerits of double kilns will be more fully discussed later on, meanwhile it will be sufficient to point out that where they are used the area reserved for kiln space or the ground plan of the maltings will be considerably less than that necessary where it is intended to use the more common type of single-floor kiln.

Malt Storage— The size of the malt stores must depend, as has been shown, to a very considerable extent on the use made of the space beneath the hot-air chamber and the size of the fire-shaft. In high kilns, where the space from the fire-chamber to the kiln floor may be from 35 to 40 feet, and where narrow shafts are used, it may be possible to store half a season's make in these chambers. If plenty of air space is left between the fire-shaft and the false shaft, and the air in this space is allowed to escape into the kiln, there should not be much danger of the malt retaining too much heat; if sufficient height is left in the stokehole to insert a hopper in one corner of it, and the malt in the store is dropped from time to time into this hopper from which it can be elevated, screened, and stored in bins reserved for screened malt, there should be practically no danger of prolonged curing.

Fig. 7 shows a section of one of these lofty, narrow shaft-kilns, and is drawn to scale to show the capacity of malt storage—the kiln is 30 feet wide and would be 35 feet long, and the height from floor to drying floor is 40 feet; the outside area of the outer shaft would be 150 square feet, giving a fire-bar area of 15 square feet, which should be ample for a 50-quarter kiln; 12 feet of head room is left in the firehole, this giving room for a hopper connecting with malt elevators to the screen and bins and capable of being filled either from the malt store, which has a capacity of some 16,000 cubic feet, or direct from the kiln. In this kiln, the size of which allows of drying some 50 quarters every five days, the malt storage provided in the kiln store would probably be sufficient. It is seldom desirable to provide storage room for less than half the quantity to be made in a season. Taking the case, then, of a 10,000-quarter house, storage for at least 5000 quarters should be provided, and, remembering that 200 quarters will be steeped at one time, it will be found convenient to make each

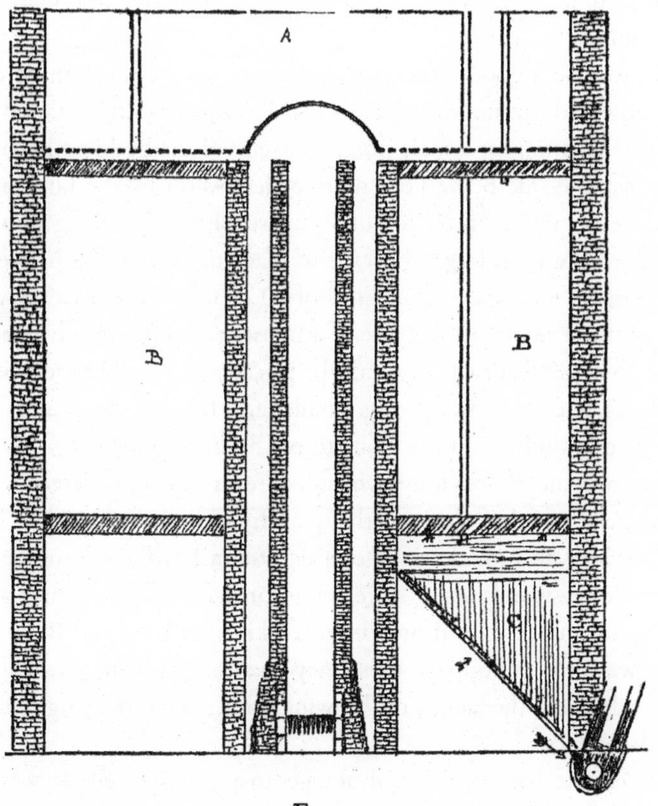

Fig. 7.

bin capable of holding either 200 or 400 quarters of malt. For it is not desirable to store a greater quantity than the latter or at most 500 quarters in any one bin, owing to the fact that the malt in any bin is most likely to get slack during the time that bin is being emptied, and that consequently the larger the bin, the longer will be the time taken in using it and the greater the danger of slackness on the surface exposed. If malt is stored in well-constructed bins, and these are trimmed up so that the air space between the top layer of malt and the top of the bin is reduced to the minimum, the moisture absorbed during twelve months will not exceed one percent, and in many cases will be much less; but supposing the bin is left half full for the same period, 2 or 3 percent may be absorbed and the upper layer rendered quite unfit for use.

Of course a good deal must depend upon the position of the malt store and the amount of outside wall forming part of it. A very favourite position for a malt store is the end of the building beyond the kilns, one side being kept more or less warm by the kiln fires, and the other taking its chance of drying sunshine or prolonged rain. But another way in large houses is to place the malt store between the kilns; for instance, if four kilns of 50 quarters capacity each are wanted, these being, in the case of a house steeping every five days, of say 35 by 30 feet each, two may be placed at the end of the working floors, the malt store being parallel and beyond these, and the other two again beyond the malt store. This arrangement will give a malt store some 70 feet long, whilst its breadth must be determined by the height of the kilns and the quantity of malt for which it has been decided to provide storage. This system has the advantage of the two longest walls of the store being constantly kept warm; it also has the advantage that, if broad-shafted kilns are used and the malt is thrown off into hoppers, these may be arranged along the length of the kilns, and the labour of throwing off very materially lightened in consequence.

To counterbalance these advantages to a certain extent, however, the two outside kilns will be a long way from the growing floors, and it will be necessary either to take the loading elevators up at an angle or to a great height to command all the kilns, unless a conveying band is used to carry the green grain to the kilns furthest removed from the growing floors, the latter being a somewhat costly piece of machinery. There is also the disadvantage, if furnace kilns are used, that the kilns will have to face different ways, and, consequently, two will generally have the benefit of the wind and dry quicker than the other two.

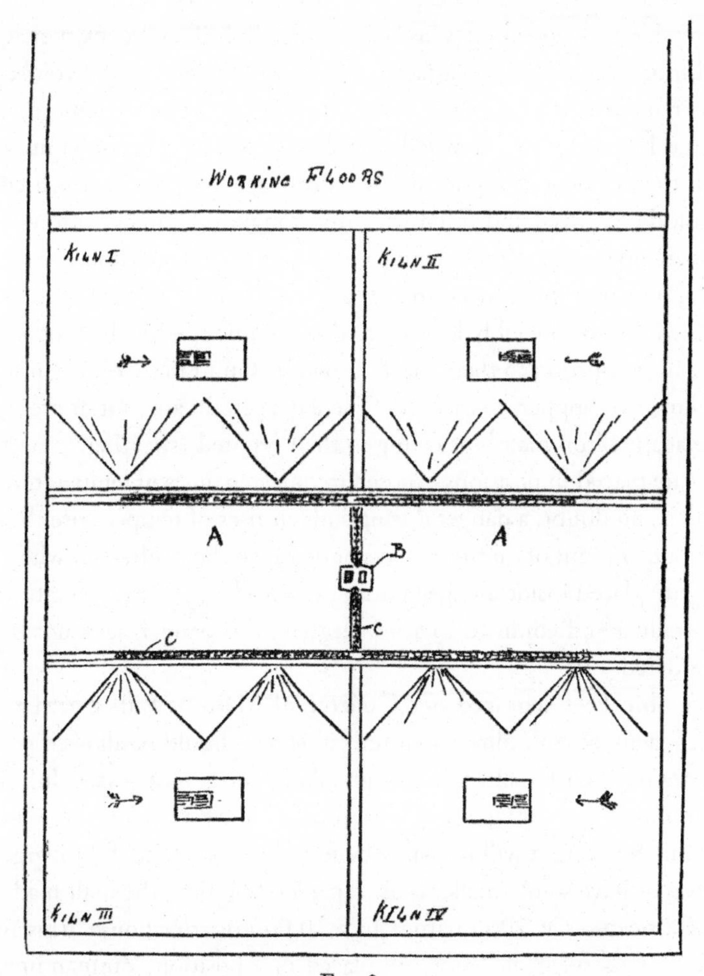

FIG. 8.

Fig. 8 represents the ground plan of construction of such kilns and malt store and will make the draught difficulty plainer, and the draught difficulty must occur to a certain extent in all furnace kilns, except where forced draught is used or the kilns are very lofty.

To turn to the bins themselves, these are best placed so as to be commanded by a conveying belt running from the malt dresser, and in deep bins a cat-ladder should be placed in each, but beyond this the sides should be as free as possible from projections. Iron bins are preferable to wooden ones, as they cannot leak if well constructed, and moreover, they resist attacks of vermin; but they are, of course, rather more costly to erect. If wooden bins are used extreme care should be taken that the wood is well seasoned, as constant contact with the dry malt is bound to cause shrinkage, and the boards used should be grooved and fitted with iron tongues. Hopper-bottomed bins are preferable to those with flat bottoms, both for ease in emptying and because the surface area always left when a bin is being emptied will be considerably less where there is a hopper; if the bins are long in proportion to their width, it will be found more economical to allow two hoppers to each bin. There is a certain amount of prejudice against iron malt bins owing to their reputed "sweating," and if they are placed in positions exposed to quick changes of temperature there is, no doubt, a danger during sudden rises of temperature that a certain amount of moisture may condense on their sides; but where they are placed inside a large building where changes of temperature are gradual and confined to a few degrees, this danger need not be considered.

Cubic Feet required per Quarter of Malt— With regard to the capacity of malt bins, about ten cubic feet should be allowed per quarter of English malt, thin foreign taking rather more space. It has been hinted that 500 quarters is as much as it is desirable to store in any one bin, and it will probably be found extremely useful in large houses to have some smaller bins, capable of holding the malt made in one steep—say, 200 quarters in a 10,000-quarter house. This is especially useful if such bins are placed in a position commanding the malt elevators, and the latter are arranged, as can be easily done in most cases, to communicate with a pipe leading back into one or more of the kilns; for even with the most careful attention to firing it may happen, especially at the commencement of a season or when a new lot of foreign barley is begun, that the curing given is either too severe or not severe enough. In the former case it will be found extremely useful to be able to keep the over-cured malt by itself, so that it can be blended into deliveries in small proportions, entirely obviating the danger, which must arise if it is run into a partly filled bin, of sending out deliveries of over-cured malt. In the second case it will be possible as occasion serves to re-dry it and bring the diastatic capacity to the point required. Entire failures do sometimes occur, though they are seldom referred to, and in the event of a 50-quarter lot of barley, bought perhaps as an experiment, going wrong, it may be extremely annoying to have to store it in a bin capable of holding 400 quarters, especially where storage room is somewhat limited.

Malt Screening— The malt-screening machinery is best placed at the top of the malt store, commanded by the elevators which bring the finished malt from the kilns, and, if it be placed as nearly as possible in the centre of the malt store, and placed high enough, it should be possible to command all the malt bins by pipes leading down from the discharger; and if it is possible to make room for an automatic malt weigher, this will be found of great advantage in estimating the amount of malt made from a given amount of barley. If, however, it is necessary to place the screen at one end of the store, it may be necessary to put in a conveying belt to carry the screened malt to the different bins.

Most modern malt screens consist of wire cylinders which rotate slowly, the culms falling through between the wires, and if the screen is at the top of the building it is as well, when possible, to leave the space directly beneath the cylinder as a store for the culms; this arrangement takes up a certain amount of space which would otherwise be malt-bin room, and entails putting up malt and culms on the same floor. It is difficult to find any objection to this, but the alternative arrangement is to sack the culms on the floor of the screening chamber and lower them by means of a sack hoist when they are sent out.

Dust Destruction— Apart from the culms separated by the screen a quantity of dust is always taken out of malt, and, if polishers are used on the screen, the amount may be considerable. This dust is lighter than the culms, and is blown out by means of a fan, and it is often a somewhat difficult question to know what to do with it; the most general way is to allow the fan to blow it into a chamber constructed for its reception with a louvered window at the end furthest removed from the inlet; if this chamber is very large most of the dust will settle before reaching the outlet and the amount blown out will be insufficient to do any damage unless the brewery is in very close proximity. But if the chamber is small a great deal will be blown out. It is difficult to prevent this because the area of the outlet should be at least equal to the inlet area and any restriction of this tends to check the draught: where it is imperative that no dust should escape, the best way, though a somewhat costly one, is to spray the outlet with water, converting the dust into sludge, which is collected and buried at the intervals necessary to keep the outlet clear. Another method, though not such an efficacious one as the former, is to stretch canvas screens across the dust chamber, these tending to filter the dust out of the air before its escape; these are satisfactory in some cases, but tend to reduce the velocity of the draught. Yet another way is to substitute a cyclone dust collector for the dust chamber, and in some cases this is the most convenient method, though it, also, has a tendency to check the velocity of the draught unless very carefully installed. Probably the most efficient way of all is to let the fan of the malt dresser blow into a cyclone dust collector, which will take out the heavier part of the dust, and let the exit from this cyclone be into a dust sludger, by means of which the lighter dust will be held in suspension in water and run down the drain in that state, when the air will escape damp but quite free from dust.

Large substances such as stones, pieces of chalk, &c., should be dealt with by the barley screen, but a few of these generally come through with the malt and are easily dealt with by riddles placed for the purpose in the malt screen; but it is impossible to take out small stones in the malt dresser, and these must be taken out by the barley dresser. Weevil, if present, are generally taken out with the malt culms, and, though sometimes a few get through into the malt bins, they seldom do any harm unless the bins are damp or the malt is allowed to get very slack. Great care should be taken that the malt screen selected for any house should be capable of screening the quantity of malt necessary in the time allotted; for instance, in a house steeping 200 quarters every five days it will be necessary, to avoid Sunday work for the machine man, to allow for a screen capable of dealing with 50 quarters a day, or, say, eight quarters an hour. But it will be much better in practice to put in malt elevators, screen, and automatic weigher, if one is used, capable of dealing with 15 quarters an hour in order to get the work over in two or three days and leave ample time for cleaning the machinery, filling grease cups, taking up belts, &c., and the increased initial outlay will be well repaid in saving of power and wear and tear.

CHAPTER II

BARLEY STORAGE

No law can be laid down for the amount of space to be set aside for the storage of barley in maltings, for it must depend upon a variety of circumstances. If the malt house is situated in a barley country and for the production of English malt only, storage capacity for one third of the season's make may be ample; but even in this case matters will depend greatly upon the season, and whether the maltster believes in sweating or not. Some are of opinion that all English barley, in no matter what condition it comes to hand, is improved by a preparatory sweat on the kiln; others prefer to sweat nothing, unless they are forced to do so by the presence of excessive moisture, while others, again, steer a medium course and sweat any barley with a moisture percentage of, say, over 16 percent.

Barley is generally better for a month's rest between sweating and steeping, and if all barley is to be sweated, proportionately more storage accommodation will be necessary. Again, many maltsters, especially brewer maltsters, prefer to buy their barley from hand to mouth and take the chance of the markets, while others prefer to buy heavily when they consider the market at its lowest. Furthermore, many sale maltsters dare not buy heavily till their malt contracts are made, and in seasons where, from one cause or another, brewers are late in making their malt contracts, heavy demands and a consequent strengthening of prices often take place towards the end of November.

Cubic Feet necessary per Quarter of Barley— In houses intended to malt foreign barley only, much depends upon whether it is intended to buy cargoes or parcels; if the former, storage for a whole season's make will generally be required, unless advantage can be taken of warehouses in the neighbourhood where warehouse rates are moderate. If it is intended to buy parcels, much must depend upon the buyer and the storage rates in the port from which the supply is to be drawn—a good general rule for anything but cargoes being to provide storage for half the quantity to be malted in a season.

Barley Bins— Ten cubic feet should be allowed per quarter of barley, so that it will be seen that to store 10,000 quarters a very large space will be required. It is also essential that barley should be stored in perfectly dry, clean, cool, and well-ventilated stores, the surfaces being as free as possible from excrescences or openings which would harbour dirt, vermin, or weevil. Wooden bins are almost universally used, and to suggest iron bins for barley would seem almost revolutionary yet there is much to be said for them. It would be practically impossible for mice and rats to make their runs through them, and between them and the walls of the malt house, as they almost invariably do in the cases of wooden bins; and whereas with the best constructed wooden bins spaces must, in time, be formed between the planks, even where these are grooved and tongued with wood or iron, no such crannies could exist in well constructed iron bins.

In many mills and large granaries or grain elevators the bins are constructed of somewhat narrow planks laid upon each other, the walls thus formed presenting an even and practically impermeable surface; but the weight and the cost of bins built up like this place them beyond the reach of the practical maltster. Again, while the danger of iron bins sweating may be disregarded in the malt store placed in close proximity to the malt kilns and kept at fairly even temperature by its contents, it can hardly be dismissed so lightly in the barley store, which is often placed above the top working floor and is far more subject to changes of temperature.

Top Storage— The advantages of having the barley store at the top of the maltings are twofold: (1) The store acts as a screen between the top working floor and the roof, protecting the latter from heat in autumn and spring and cold in winter; and (2) the barley, having been stored at the top, is in a position to gravitate down to the screen, sweating kiln, or steep. When barley is to be stored in this position it is probably best to make as few bins as possible; but it is not at all certain that this top storage for large quantities of barley is very desirable.

In order to explain these disadvantages it will be best to refer again to a 10,000-quarter house. To begin with, to carry a weight of 1000 tons-the weight of 5000 quarters of barley—at the top of the house will entail building a very solid and expensive house. The area of the barley store will, of course, depend upon the number of working floors allowed, and, as it is practically impossible that there will be less than four, the maximum superficial area will be about 8750 square feet, say 70 by 125; it will be necessary to cut off some 20-25 feet of the length for screening plant and steeping cisterns, and of the width at least six feet will be taken by a gangway, left preferably down the middle. This will leave an area of some 64 by 100, say 6500 square feet; so that to hold the full 5000 quarters, allowing 10 cubic feet to the quarter, it will be necessary to store the barley to a depth of seven or eight feet.

These modern top-storage lofts are generally filled by a conveying belt fed by the intake elevators and running the length of the store above the gangway, from which the barley can be shot off at any desired point on to the floor below and on either side of the gangway; the latter divides the store into two equal parts, and in houses destined for making up only brewing foreign it may be necessary to make no divisions. Where, however, English and foreign have to be malted, it will probably be necessary to divide each side into bins in order to separate any different varieties that may be bought, and to keep the dried English separate from the undried. Whether the store is divided into separate compartments or not, the process of conveying the barley to the steep or to the screens before steeping will have to be done by manual labour, and will be a somewhat long job especially when the barley furthest from the steeps has to be dealt with. It will probably keep two men busy, in the case of screening for the time necessary according to the capacity of the screen and the cleanliness of the barley.

When provision has to be made for sweating English barley where the barley store is at the top, either one or two bins must be set aside for the storage of the barley to be sweated, and these bins should be near the intake elevators, which should also be commanded by the sweating kiln or drum. Where all or a proportion of brewing foreign has to be malted, provision should be made for taking out as much as possible of the dust and dirt when it first comes into the house, and this can be conveniently accomplished by placing a "rougher-out," capable of doing its work as fast as the elevators between the discharge and the conveying belt. It is a mistake to let the only discharge be through the rougher-out, for sometimes it is not necessary and its use will not improve the barley. Any unnecessary contact with machinery tends to skin and break barleys, especially when they are dry, so that two-way spout must be used, one arm leading straight to the belt and the other to the rougher-out. If above this again an automatic weigher is placed, the elevator head will have to be carried to a corresponding height above the barley store, and this, of course, will mean increased initial expenditure, which the increased convenience in working will very amply repay. It is the greatest mistake to skimp the height or the depth of elevators, and, though it is often done for the sake of saving a few pounds in cost, it is invariably bad policy.

The barley intake should be at the cistern end of the building and removed as far as possible from the malt kilns; and the whole of the barley screening plant and, if possible, the steeping cisterns also should be cut off from the working floors, any doors connecting one part of the building with the other being arranged to close automatically.

Although some of the drawbacks to top storage have been pointed out, the fact remains that there must be some shield between the top working floor and the roof, and that a somewhat lofty chamber such as a barley store makes the best possible shield from heat or cold. Partly to protect the barley from variations of temperature a false roof of wood is generally built in some foot or so below the slates or tiles, but even then it is often dangerous to let the barley lie at any depth in the store throughout a hot summer without very careful watching. If the garret, as it was called in old houses, be not used as a barley store it becomes so much wasted space, and it is therefore usual to store some at least of the barley there, the rest being stored in bins built at the end of the house far removed from the kilns. These bins are generally in line at the end of the house, the sweating kiln or drum being immediately behind them, or at one end in such a position as not to interfere with draught for the working floors; they may be of great depth, equal to the height of the house less head room beneath and above them. Supposing a house 70 feet broad, seven bins nearly 10 feet in breadth and 20 to 25 feet in height could be conveniently placed, their length depending upon the amount of barley to be stored.

Beneath these bins, and running across the breadth of the house, it is customary to have a band communicating with the intake elevators, which may either be placed in the middle of the building (in which case the belt serves as an intake and must be reversible) or may be taken up at an angle from the end where the barley is shot into them to the centre of the building at its highest point. In this latter case the conveying band beneath the bins need not be reversible, its only function being to convey the barley stored in the deep bins to the intake elevators for screening and steeping, or sometimes in cases of English barley to the sweating kiln; but it is not advisable to store unsweated English barley for long periods in these deep bins unless the moisture percentage in it is very low, say 13 to 14 percent. As a matter of general principle, end storage is more suitable for foreign brewing barleys than English; the drawbacks to it are the great depth of the bins and the fact that all barley has to be elevated twice—once on coming in and again before it is steeped. For English barley, where it is necessary to sweat, it is much better to leave storage above the top working floor so that it may gravitate down to the kiln or drum.

Wherever the barley is stored it is advisable to have garners immediately over the steeping cisterns capable of holding one or two wettings for each, and if, as is probably most convenient, the cisterns are placed above the working floor so as to empty practically automatically, these garners will have to be carried to a considerable height. Hence it will be necessary to elevate the barley into them either direct from the screen or, in case of clean barley, direct from store, unless it be possible to steep direct from the intake elevators, where the latter can be arranged to command the steeping garners. In any case it is not advisable to attempt to place the barley screens and half-corn cylinders above the steeping garner; it is far better, on the contrary, to place them rather low in the houses and let them be commanded by a large hopper capable of holding sufficient barley for one steep for all the cisterns. The reason for this is that screening to be efficient must be a slow process, and if the storages in deep bins at the end of the building, necessitating the elevation of the barley to the screen by the intake elevators, capable probably of dealing with some 50 quarters an hour, a great waste of power will ensue if the feed be restricted to the pace at which the barley can be screened—12 to 15 quarters per hour. If, on the other hand, the storage bins are at the top of the building and the barley has to be filled into barrows and run to the screen, two men's time will be taken up for the time necessary, which will be much shorter if the barley has to be shot into a hopper than if they have to wait for it to be screened; in this case there is always a tendency for the screening to be got over as quickly as possible.

It has been pointed out that barley stores placed above the top working floors of maltings are generally bad summer stores by reason of the heat from the roof, and that false roofs with an air space of a foot or so afford little protection in a hot summer. If, however, these stores are divided into two floors each, say, eight feet high, with the conveying belt running in the angle of the roof immediately above the top one, and any barley to be stored during the summer months is kept in the lower store, little danger of high temperature should arise, the top floor acting as an efficient screen between the roof and the lower floor. This division will also enable English and foreign barley to be kept apart, and if gangways are left down the middle of each store, opportunity will be afforded of keeping different varieties of both English and foreign apart.

Fig. 9 will give an idea of the top-storage system. In it are represented the main intake elevators (1) delivering into the automatic· weigher (2), a spout from which delivers the barley into the rougher-out (3), from which spouts (4) (5) and (6) lead respectively into the worm (7) feeding steeping garner G, which commands the steeps

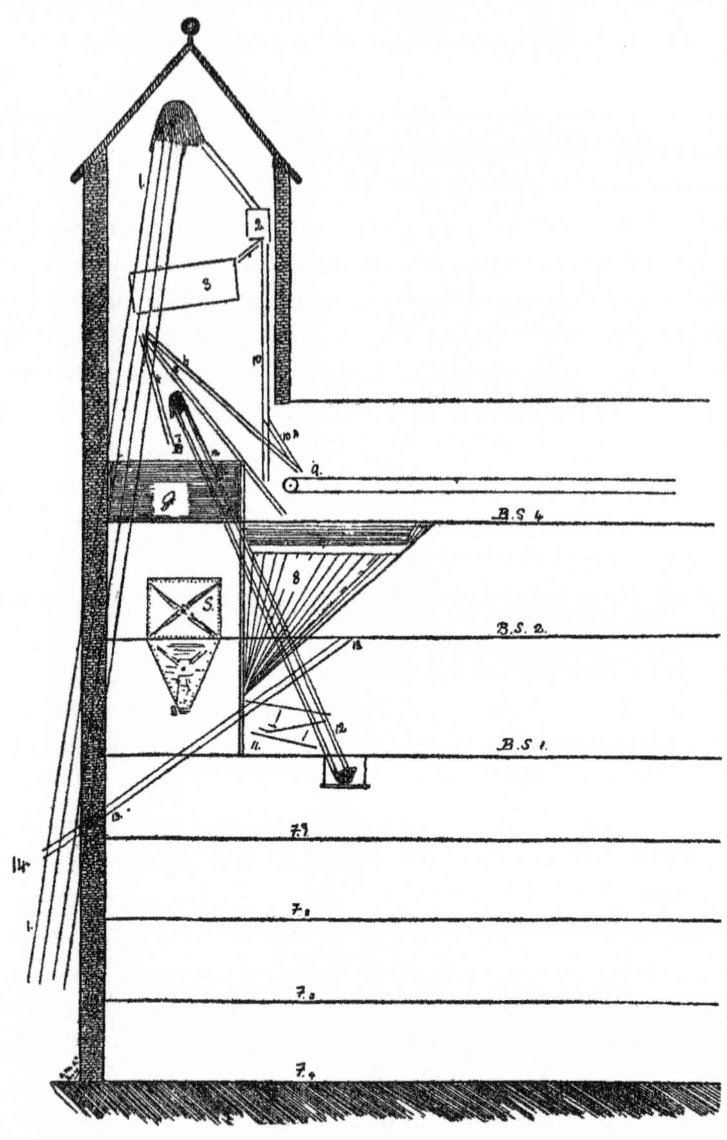

Fig. 9.

S if the barley is only to be roughly screened and steeped immediately, the conveying belt (9) which commands the two barley stores, B.S.1 and B.S.2, if the barley is to be stored, and the screening hopper (8), which commands the finishing screen and half-corn cylinders (11), if thorough screening is necessary. The latter are arranged to deliver into elevators (12), which convey the screened barley to the steeping garners, and these elevators would also probably serve to elevate the barley stored in the top and bottom stores into the screening hopper (8) when necessary. In case the barley does not need to go over the rougher-out, alternative spouts, (10) and (10a), lead respectively to the hopper (8) commanding the finishing screen and the conveying belt(9) for storage.

B.S.4 represents the garret along which the distributing belt (9) runs, and could be used to store a small quantity of barley if necessary, but, as will be seen by comparison with Fig. II, p. 51, this garret is usually very narrow and would afford little room for storage.

(13) indicates the position of a spout or spouts from the upper barley floor leading to (14), which is the approximate position of the sweating kiln or drum, the throw-off from which would command the main elevators (1), so that the dried barley could be elevated daily during sweating operations in a very short time, and stored in the lower floor till ready for use. F. 1, 2, 3, 4 are of course the growing floors commanded by the steeps. Compare Figs. 10 and 11, pp. 50, 51.

Deterioration of Barley in Store— Barley in store may deteriorate from several causes, the chief of these being

Heat,
Weevil,
Vermin.

1. *Liability of Barley to Heat in Store*— When undried English barley is stored in bulk with a high percentage of contained moisture (say over 15 percent) and remains untouched for some time, there is always danger that it may get warm.

The probability of its so doing is dependent upon the condition and the moisture of the barley when stored, the depth at which it is stored, and the cleanliness, airiness, temperature, and general suitability of the store.

Barley stores should be cool, clean, airy chambers, and it is especially necessary that the floors, walls, and partitions in them should have smooth and regular surfaces, inequalities and holes being especially to be avoided owing to their likelihood to harbour weevil, vermin, mould, &c., and of course it is of extreme importance that no dampness should be present, either owing to damp walls or faulty roofs. Where they are at the top of a building a false roof is useful in preventing too rapid changes of temperature. There should be as few skylight windows present as possible, and these should preferably be of that form of glazing which tends to prevent condensation and consequent drip as much as possible.

No barley with a moisture percentage of over 15 should be left in bulk for long without being carefully watched and sampled every few days, especially if it is stored in bins or at a depth of over three or four feet.

Many maltsters never store barley over three feet deep, but this is possibly carrying caution to an extreme point, as the saving of space by deep storage is very considerable.

If barley be stored in bins over three feet deep it is a good plan to keep several thermometers in the heap and watch temperature carefully, as a corner of a heap will sometimes start getting warm very rapidly, and this heat will soon spread through the bulk, although this does not as a rule happen in dry, well-ventilated stores unless the moisture content of the barley is very high, or weevil are present, or the temperature of the air in store is very high—70° F. or over.

When dry barley, such as dried English or the dry foreign sorts, heats in the bin the cause is almost invariably either a warm store or the presence of weevil, except in those cases in which English barley has been sweated and stored either with rather a high moisture content—12 percent and over—or successive dryings have been run into the same bin without having had time to cool, especially when it has been necessary to trim and tread down the barley.

It must be remembered that when barley gets warm a certain

amount of moisture is thrown off. This sweat is only apparent to the eye in very extreme cases, but can be sometimes recognised by handling the barley, and on the extent to which it can be carried off by air the ultimate condition of the barley must very largely depend.

When this exhaled moisture is not carried off by air the conditions in the heap are most conducive to heat, warmth, and moisture without much air being present.

The heating process, therefore, is in part a process of suffocation, and must be met by aeration, and for this reason it is most imperative that directly a heap shows any sign of becoming warm it should be moved and aerated as much as possible.

This is usually done in practice by turning the barley with the shovel, and of course it is upon the depth of the heap that the ease with which the barley can be turned depends, so that where there is any danger of heat, either owing to weevil, summer storage, high moisture content, or partial sweating, it is most inadvisable that barley should be stored in heavy bulk.

Where, however, the barley is stored perfectly dry and cool in suitable stores, or where it is stored for a short time only, the chance of heat is so remote that there is practically no risk in storing in bulk, though this should never be done if it can possibly be avoided where either the barley has to be in store for a long time, say during the summer months, or where weevil are present in the house.

2. *Weevil* are introduced into this country in foreign barley, chiefly in Brewing Californian.

They are small insects varying in size up to rather under ⅙ in. long, with long hard beaks. In colour they change with age from a red-brown to darker brown or black.

There are often two generations in a summer, the first being born in April and the second in September.

The pregnant female deposits her eggs one in each corn in a hole which she hollows out for the purpose usually in the ventral furrow. The larvae hatch out in about a fortnight and immediately proceed to gnaw through the contents of the corn, from the empty husk of which the adult insects proceed in about ten or eleven weeks.

It has been estimated that between 5000 and 10,000 insects proceed from a pair in one season under favourable conditions[1]: these are a warm, moist atmosphere and freedom from draught.

It is apparent that if weevil are present in barley and it is carefully screened *before* the breeding stage most of the insects can be collected and burnt. Once the eggs are laid, however, screening is practically useless, the larvae-containing corns being inseparable from the others.

If the barley can be shut up in an atmosphere of carbon bisulphide gas for ten or twelve hours both the adult weevil and the larvae will be destroyed. Carbon bisulphide is a very inflammable gas, and its use cannot be recommended as a rule.

Another method consists of pouring dry sand into the heap till the air space is filled up. That this is efficient is proved by the refusal of weevil to go in to those barleys which normally contain a good deal of sand, such as Syrian, Tripoli, and North Africans, but it does not seem very practical.

Practically the only way to deal with weeviled barley in bulk is to spread it out as thin as possible and to keep it turned regularly, or, better still, to put a good depth of it on to an empty kiln where the draught will prevent the spread of the insects and eventually drive them off or kill them.

Both these courses can be adopted only in summer, as a rule, when the floors and kilns are empty, but of course it is at this time of the year when weevil are most likely to be prevalent.

If weeviled barley is stored in bulk it is very likely to get warm. When this happens and it is impossible to spread it out, then the best plan is to test it for germination, and, if this is satisfactory, screen and steep it as soon as possible; the weevil will not be killed off either during steeping, flooring, or kilning, but will do no damage in the malt bin unless that is damp or the malt is very slack, and of course almost all of them will probably be taken out by the malt screen.

Moth are sometimes present in South European and North African barley, but as far as is known they do not breed in this country, and the damage done by them is done before the arrival of the barley.

---

1 *Petit Journal de Brasseur*, 1903, p. 1029, ref. M. Poskin, Prof. Stats. fost. Agriculture, Belgium;Brewing Trade Review, pp. 17 and 85, vol. xix. 1905.

Moth-eaten corns are often very difficult to get rid of, as they are frequently only partly eaten and consequently too heavy to blow out or swim out. If a large proportion is present the malt will therefore be spoilt, as the moth-eaten corns are sure to mould on the floors.

Mice and rats often do a great deal of damage both in barley and malt stores, the latter especially on waterside maltings. It is difficult to find a remedy for mice, but rats can be kept out to a certain extent by fitting wirecovers over the drain pipes by which they often make their way into the malt-house.

CHAPTER III

MACHINERY

FROM the economical standpoint, machinery plays an important part in modern malt-house construction, and the saving in working expenses which can be effected by such machinery must vary very considerably according to the knowledge of working requirements brought to bear both upon the construction of the buildings themselves and the placing of the various conveyors, screens, and other plant with regard to their ultimate capacity for, and adaptability to work in the finished building. It must be remembered that machinery, such as sack hoists or elevators for taking in barley or loading green grain is not only convenient but absolutely necessary in high modern buildings, and that to say that the cost of labour in these modern houses, plus interest and depreciation on machinery, should equal labour alone in small old-fashioned houses is scarcely fair to the modern houses where, apart altogether from the cost of labour and interest and depreciation on machinery, there is, or should be, a saving owing to increased convenience in getting barley and malt to and from the malt-house.

In his excellent treatise on "The Flooring and Kilning of Malt," published by the *Brewing Trade Review* several years ago, Mr. THOS. HAYNES, jun., of Radcliffe-on-Trent, called attention to the advisability of having the drying kiln and the cistern at the same end of small malt-houses. In this type of house the cistern is placed on the ground floor, and the green malt is worked to the far end of the bottom floor by about the fifth day out, when it is pulled up to the top floor and worked back to the kiln, whose floor is generally placed a foot or two below the top floor for convenience in loading the green malt which has to be carried on, the intermediate floor serving as a store for barley and malt. When the kiln is used at the commencement of the season for sweating barley, it is very convenient for the latter to be thrown off on to the floor immediately above the steeping cisterns, whereas if the cisterns are at the end of the building, far removed from the kiln, it has to be carried back the entire length of the building. This may serve as an example of the way constructional detail may influence labour in houses where no machinery is used, but the case becomes very much more complex where the building has to be constructed for the economical arrangement of machinery.

Power in malt-houses generally comes from a gas engine or electric motors, steam being very seldom used. Where the Board of Trade unit is cheap, or where several large houses standing in proximity to one another justify the erection of a suction gas plant for generating electric energy, motors are far the most convenient, and render long belt drives and lengths of shafting unnecessary; but where the Board of Trade unit costs *2½d.* or more, a certain saving will be effected by using a gas engine. Where suction gas plant is installed for groups of houses due allowance must be made for somewhat heavy depreciation on the total plant in addition to the depreciation on motors. One man should easily look after a suction gas plant of sufficient size for three 10,000-quarter houses, and it is of course advisable to have two gas engines in case of breakdown in large plants. It is probably inadvisable to install gas suction plants where the Board of Trade unit can be had for *1½d.* or under.

For the intake of barley much must depend upon circumstances. Where the maltings are situated on a river or a canal and a great deal of foreign barley is to be taken in from barges, into which it has been bulked ex ship, it may be advisable to erect the type of elevators known as a "sea-leg," which can be lowered into the barge from the wharf, and discharge the barley into a belt leading into the house, or, where the house is built right on the water, straight into the main intake elevators. Even this must depend upon the type of barley which is to be worked. Brewing Californian barley, for instance, is imported in cental or 100-lb. bags, and it is customary on loading the ship at San Francisco to cut open a great many of these bags in order to save room on the ship, so that on arrival in this country a considerable proportion of the barley actually arrives in bulk. When the ship is unloaded either all the barley is weighed in original bags and shot in bulk into the barges, or loaded in original bags into barge, the contents of the torn bags being shot in loose. On the other hand, the barley may be shot out of the original bags into English sacks and weighed into barge at even weights, this method of course entailing a charge for sack hire and some extra charge for working. Chilian barley, on the contrary is imported in stronger bags, holding about 200-220 lb., and these bags can be handled almost as easily as English sacks, so that there would be little advantage in the case of Chilian barley in using the sea-leg; and if this is not used, a small crane on the wharf, capable of lifting three or four sacks together, will save a great deal of labour.

Where the barley is to be taken in from farmers' carts or railway trucks, elevators or endless conveying belts are more suitable than the more old-fashioned sack hoist, as they can deal with large quantities much faster and with far less manual labor. The only disadvantage about elevators is that the barley must be carefully sampled before being shot into them, as once the grain is stored in bulk it cannot be returned, while if stored in sacks and found inferior to sample it can be kept apart till the necessary arrangements with regard to it have been made. For this reason it is advisable where possible to have a covered shed capable of holding some 50 quarters of barley in sack adjacent to the elevator intake, and, if sweating accommodation be installed, the ground floor of the kiln or drum shed will answer this purpose admirably. According to the arrangements made for storage, the intake will be on to a belt, feeding vertical elevators, or direct into elevators set at an angle, bringing their head or discharge into a central position at the top of the building. The latter arrangement is preferable where possible, because as a matter of principle conveying belts should be made to run one way only. Careful adjustment is necessary with these belts, in order that they should run in the centre of the pulleys, and where it is necessary to reverse them the adjustment becomes a much more difficult matter than when they are made to run always in the same direction. When they do not run true there is chafing against the sides, and in time the sides of the belt get worn and fray belts are made either of canvas or canvas covered with india-rubber, and when thy fray there is danger that the frayed strands of canvas will catch round projections, such as grease-pots, &c., while running, and rip off portions of the sides of the belt.

Whether the barley is to be shot on to a belt or into elevators, sufficient height should be left to allow for a small hopper capable of holding three or four sacks, and this hopper should be kept full as much as possible to ensure a steady feed, this being adjusted by means of an iron slide from the hopper to the belt or elevators. This is especially necessary in the case of belts, as an uneven feed tends to cause the grain to scatter off the belt instead of running smoothly on to it. Elevators are driven from the top, but plenty of room should be left at the bottom to clear out the barley in case of "chokes," and if the bottoms are sunk low into the ground on a badly draining soil, or where there is any chance of flooding, they should be placed in a cast-iron tank instead of the ordinary brick or concrete chamber. Elevators, for barley and malt consist of iron cups fastened to endless belts either of canvas or rubber, but in those for green grain the cups are preferably carried by chains, these obviating the danger of the green malt being crushed between the belt and top pulley. The cases are made either of pitch pine or iron, and, though pitch pine is the commoner material, there is much to be said for iron pipes; this is especially the case at the discharge, where two- or three-way arms are necessary, for the wood has to be lifted with metal to prevent its being worn out by the constant friction, even with green malt, of the discharges of material. Moreover, between the wood and the lining corns find their way, so that the valves are apt to get jammed and not to shut close, causing a good deal of inconvenience and constant repairing, which can be dispensed with by the use of iron cases and spouts.

Roughers-out for Foreign Barley— Where an automatic weigher is placed to intercept the discharge from the elevator head, there should be a two-way spout, one way leading to and the other avoiding the weigher. Beneath it again it may be necessary to have a two- or three-way spout leading either through a rougher-out or avoiding it, or leading to the steeping garners for clean barley or to the storage bins, top belt, or sweating kilns, as may be needed for different classes of material. Dry English barley will run at an angle of 30°, but it is necessary to give rather more fall where damp English and thin foreign have to be allowed for. Where the intake elevators are to be taken high it is generally as well to take the last 15 or 20 feet up in a tower. If the rougher-out is placed in this tower it will occupy some 8 or 10 feet; it will consist of a wire cylinder some 3 or 4 feet in diameter and 5 or 6 feet long, the wires being placed very close together at the part where the barley enters to riddle out any small stones, sand, &c., and much wider apart about half-way down the remainder of the length in order to allow the barley to fall through, while large substances, such as straws, large stones, string, &c., are kept back and fall out at the end of the cylinder; this cylinder is set at a slight angle to the floor and a worm of diameter equal to the inside of the cylinder is fixed to it, which rotates with the cylinder and carries the barley forward. It is customary in construction greatly to increase the number of turns in this worm as it approaches the part of the cylinder through which the barley falls, in order to arrest the feed as much as possible. The barley falls into a hopper below the second part of the cylinder, and usually a fan is allowed to play on it as it falls, blowing out light dust into the dust chamber or sludger, while the heavier dust is trapped and caught in a sack fixed for that purpose at the exit.

The principle on which these traps are contructed is an increase in the size of the air shaft, by which the velocity of the current of air is reduced and heavy particles consequently dropped, the decreased velocity being sufficient to carry on the light dust only. Roughers-out may be obtained of sufficient size to deal with 50-60 quarters an hour, or as fast as the intake elevators can work, and they require somewhat frequent attention if there is much string (sack ties, &c.) in the barley, as these tend to foul the worm where its coils are close together. Where a rougher-out is used, therefore, great care should be taken that it is of ample size. It is impossible to foresee what barleys may be bought in a succession of seasons, and while in one year barleys such as Smyrnas may come in fairly clean and the maltster may congratulate himself in having a machine capable of taking out dirt sufficiently fast, the next season may be a bad one, and owing to a wet harvest and a consequent prolific growth of weeds, tares, &c., the rougher-out may not be able to deal with them at a pace sufficient to keep elevators and men busy. Besides, there is always the chance that when barleys come to hand fairly clean, and without many broken corns, treatment by the rougher-out, if the latter be a good one and of sufficient size to do its work thoroughly, may obviate the necessity of passing the barley over the finishing screen and half-corn cylinders before steeping, thereby saving power, and—a somewhat important point—saving the skin of the barley from being unnecessarily knocked about. There is no doubt that useful and economical as machinery is, there is a danger, especially with delicate and thin-skinned barleys, that too much knocking about in elevators, worms, &c., may damage the husk and lead to danger from mould on the floors. Worms are especially bad in this way, and, though it is almost impossible to do without them altogether, they should not be used more than necessary, endless band conveyors doing far less damage.

Of course, this does not apply to malt off the kiln, where worms are rather helpful than otherwise in breaking off the dry rootlets.

Shaking Screens— Although the cylinder type of barley screen is very useful as a rougher-out, it has great disadvantages as a finishing screen. For finishing shaking beds are better, as two or even three beds of different gauge can be kept and placed in the screen according to the amount of screenings which it is desirable to take out. A riddle is generally placed first to take out large substances, and two beds of different gauge are fixed beneath it, making two grades of screenings and keeping dust, dirt, and large substances separate. With the cylinder type of finishing screen the dust and dirt are generally taken out with the screenings, which have either to be screened again before being sold, or sold at a much lower price than would be obtained if they were comparatively clean.

Half-corn Cylinders— It is always advisable to put in efficient machinery for taking out half-corns, as these if left in the barley, are a prolific source of mould on the floors. This machinery generally consists of one or more indented cylinders of slow rotation, the corns falling into the indentations; as the cylinder revolves upward its surface is brought into contact with a scraper knocking out the projecting whole corns, which thus cannot rise above it, and are carried along the bottom of the cylinder to the exit; the half-corns, which escape the scraper, are carried to the top of the cylinder, from which they fall into a worm running down the length of the cylinder, which is contained in a trough, one of whose sides forms the scraper referred to above. It is possible to adjust the scraper nearer to or further from the inside surface of the cylinder, in order to accommodate barleys of varying length, a round cobby English Chevallier requiring it to be set much nearer the surface than a long thin brewing barley such as Brewing Chilian or Yerli Smyrna. From the half-corn cylinders the barley will generally have to be elevated into the garner or garners above the steeping cisterns, and at the dis charge of these elevators, also, it will be found useful to have an automatic weigher, so as to be sure that the right quantity of barley is steeped, and to be able to tell the amount of substance wasted, generally referred to as malting loss, on each steep.

Aeration in Steep— The cisterns themselves can hardly be classed as machinery, but, as their construction and arrangement have considerable effect upon labour, it may be as well briefly to refer to them here. A good many theories have been put forward lately, and in Germany a certain amount of practical work has been done on the effects of aeration in steep, and, in order thoroughly to aerate the grain during steeping, arrangements have been made for blowing air through the grain during steeping, and even blowing grain and water from one cistern to another. There can be little doubt that a certain amount of aeration well applied does stimulate germinative energy, but it is improbable that the time has yet arrived when sufficient is known to justify much money being spent upon pumping grain and air.

A certain amount of aeration can always be effected by using sparge arms over the cisterns, when each drop of water will carry a certain amount of air down with it, and by draining the cisterns from the bottom, especially with hopper bottoms, when the water is changed, so that the retreating water will suck air after it into the interstices between the corns. Besides the sparge arms, underlets should be supplied as the inrush of water will tend to lighten up the grain after it has been sucked down, and to a certain extent compressed by draining.

Self-emptying Cisterns— The mistake most often made with hopper-bottomed cisterns is to let the exit be too near the floor into which the barley is to be discharged from steep. Malting floors generally take up seven feet from the surface of one to the surface of the one above or below, seldom or never more than eight feet, and the discharge of any cistern should be some two feet above the surface of the floor above that on to which the barley is to be discharged, when all that will be necessary for emptying will be to open the valve, having given sufficient time for efficient drainage, and let the barley run out, the sides of the cistern being brushed down afterwards. If, however, the exit of the hopper is only three or four feet above the floor on which the steeped barley is to lie, it will be necessary to keep two men with shovels trimming the steeped grain away as it comes out of the hopper, making emptying a longer process and one requiring considerably more labour. It is a mistake, however, to have each cistern two floors above the floor corresponding, to it, as is often done, as in these cases the cisterns must occupy part of the working floors and besides taking up floor space it, will be impossible to keep barley dust off the working floors during the time the barley is run into steep, whether the water has been run in first or not. It is far better to arrange all the cisterns upon one floor commanding the top working floor, but absolutely shut off from it (Fig. 10) or on two floors (Fig. 11) so that all barley dust is excluded from the working floors; the drawback to Fig. 10 being the width taken up by four cisterns in line. In this sketch, the ground floor and the three floors above it represent working floors, the two floors above them being for barley storage; the roof of the working floors would probably spring from the top of the top floor, and a tower taken up for the cisterns, garners, and barley elevators, so that if the cisterns were in line this tower would have to be of great size.

Power Shovels— Once the steeped barley has been emptied from the cisterns, the process of turning brings it forward toward the kilns, the end of each piece furthest removed from the latter being as a rule not more than one third of the length of the floor from the loading elevator when the malt is ready for kilning, except

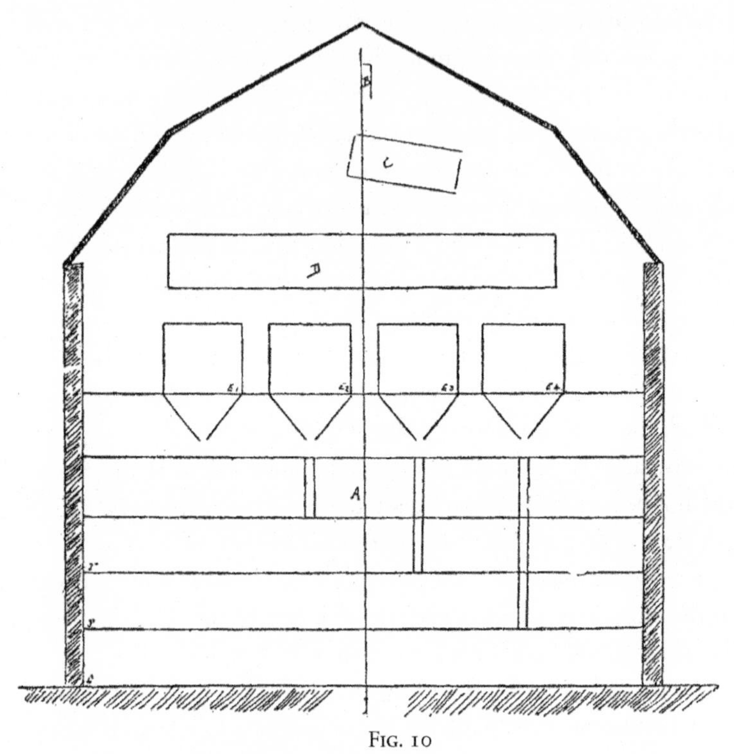

FIG. 10

A—Show position of barley elevators
B—Automatic weigher.
C—Rougher out.
D—Steeping garner.
$E_{1, 2, 3, 4}$—Steeps.
1, 2, 3, 4s—Floors corresponding to steeps.

in very warm weather. In England it is customary either to carry it in carrying boxes or wheel it on barrows to the elevators, and if the latter are used care should be taken that the wheels are very broad in order that they may be easily wheeled over the thickness of growing grain. An alternative system is to have iron rails running beneath the girders supporting the floors from which baskets running on wheels on these rails can be filled at any point and pushed forward to the

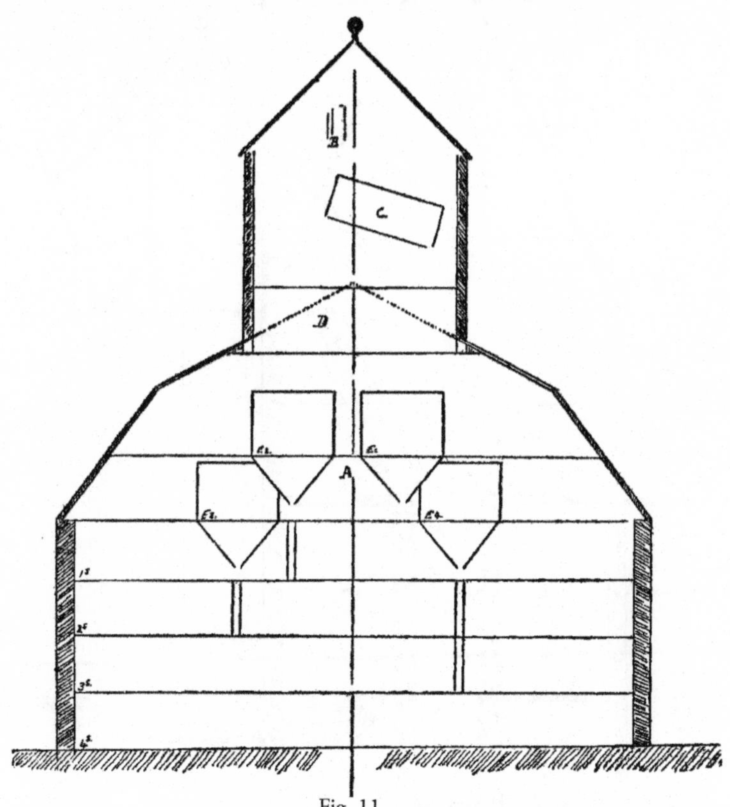

Fig. 11
A—Shows position of barley elevators.
B—Shows automatic weigher.
C—Rougher out.
D—Steeping garner.
E 1, 2, 3, 4—Steeps.
1, 2, 3, 4s—Floors corresponding to steeps.

elevators. In America there is a more general use of the power shovel for bringing the green malt to the elevators this being a large wooden or iron shovel connected by a rope or wire to a friction hoist, small

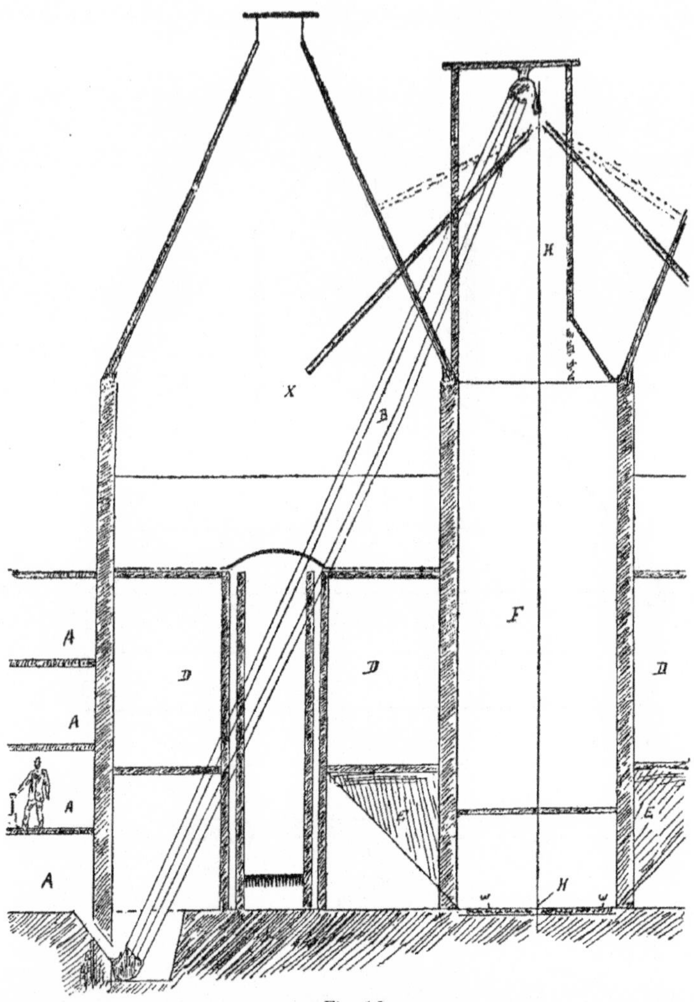

Fig. 12
In this sketch AA represent the working floors from which the elevators B take the malt for distribution in the four kilns through the spouts CC. The curve necessary on these elevators for the sag of the descending chain is shown. DD is storage for unscreened malt. E is a hopper communicating with elevators whose position is indicated by the line H though worm W. F indicates position of bins for screened malt. The swivel-spout for distribution would be fixed at X.

pulleys being used to bring the cord round corners; the power is generally taken from a pulley off the elevator drive, and it is necessary for one man to stand at the elevators to control the friction while another walks behind the shovel and directs it. There is no doubt that this arrangement saves a good deal of labour, but it is not very suitable for floors supported by pillars, as it is generally difficult to negotiate the shovel round them.

The green grain elevators are preferably chain elevators, and their bottom should be sunk well below the bottom floor so that the grain on it may not have to be lifted into the feed; for loading the malt from the other floors holes must be left in the floor immediately above the feed, and it will save some sweeping if a spout be taken down from each floor to within two feet of the floor below; where the elevators are vertical, the spout will run parallel to and alongside them and cause no inconvenience, but where they are set at an angle it may be better to dispense with it. Wherever elevators, and especially chain elevators, are set at an angle great care will have to be taken in their erection; the drive being from the top the ascending portion of the belt or chains will of course be kept tight, but the descending portion will be sure to sag over somewhat; hence great care must be taken in erecting the casing that the cups shall not grate against it when descending, and to ensure this it will probably be necessary to make it bulge out rather from the ascending shaft. Fig. 12 shows this.

Where it can be arranged to take the elevators and deliver the malt from a good height into the middle of the kiln swivel-spouts will in loading be of great advantage in saving labour; they need not be of size to interfere seriously with draught. Automatic kiln turners are now used in some malt kilns, and they are probably better for drying off than the best manual labour, if properly adjusted. There is also the advantage with them that, given an efficient night man who understands them and is a good stoker, kilns can be dried off at night, when otherwise they would have to be left till next day and a certain amount of time and fuel wasted. They are, however, somewhat expensive to install, and the labour they save is probably more than balanced by the amount of power they use, and until the grain is hand-dry very low heats have to be maintained or there may be serious danger of vitrification owing to the green malt being turned on to the kiln floor. Reference will again be made to them, but meanwhile they cannot be looked upon as economical labour-saving appliances.

CHAPTER IV

BARLEY DRYING

Height of Barley Sweating Kilns— Allusion has already been made to the desirability of arranging for drying English barleys when necessary, arid for storing barley to be dried where it can gravitate to the sweating kiln or drum. If a kiln is used, about 16 square feet should be allowed per quarter of barley to be sweated. Great height is not necessary between the fire and the drying floor, 15 feet being ample for a 50-quarter kiln. The drying floor may be of well-constructed tiles, these admitting abundant draught, or of the more costly Hermann wire. The draught holes in tiles, however, are rather apt to get choked by corns which fall into them and get broken by the shovel, so that care should be taken to clear them from time to time. Plenty of draught holes should be left in the shaft, whether a fire basket or furnace is used.

Temperatures for Sweating Barley on Kiln— When the damp barley is loaded on to the kiln, a temperature of 80° F. should not be exceeded, but the temperature, indicated by a thermometer resting on the tiles surrounded by the grain, may be brought to 120° F. when drying has been in progress three or four hours. It is not advisable to aim at saturating the air at any time, and a limit of 80 percent of saturation should not be exceeded. Whilst the moisture present in the barley is sufficient to bring the saturation point of the air near this limit, it is as well to pass a good deal of air through the grain, but as it becomes drier, the draught may be reduced, and the temperature raised to 130° F., and though barleys have been subjected to heats of 140° F. and even 160° F. without their vitality being destroyed, such heats are, to say the least of it, extremely risky, and should not be approached. As regards the time necessary for drying, much, of course, will depend upon the condition of the barley to be sweated, and the amount of moisture which is to be left in it; but allowing 16 square feet to the quarter, 24 hours will be required at the most to bring the dampest barleys to a moisture percentage of 10 to 11, during which time the barley should be turned once or twice with the shovel.

Beneath the hot-air chamber a hopper of sufficient size should be placed, communicating with elevators for the removal of the barley.

Barley Sweating Drums— In the case of sweating drums, the conditions are somewhat different. The air is sucked through the drum by a fan placed at the end of it farthest from the furnace, so that the amount of air passing through the barley is regulated to a great extent by the resistance offered by the barley.

Temperatures for Sweating Barley in Drums— The temperature is regulated by the size of the fire and by cold air admitted through draught holes placed between the furnace and the drum; it is recorded on a thermometer placed in the ·shaft, and is unaffected by evaporation, so that a recorded temperature of 140° F. is safe two or three hours after beginning drying. The drum, rotating slowly in one direction, leaves the surface of the barley at a constant angle of about 35°.

Initial cost of Kilns and Drums— The air on entering the drum passes through perforated channels going the whole length of the drum, and from these channels, of course, through the perforation in them to and through the barley, and out through a perforated centre tube connected with the fan. There is a swing valve arranged so that no air can enter such of the inlet arms as are at any time above the level of the barley. Valves are left in the surface of the drum for admitting and letting out the barley, and the drum has to be rotated till these are in position, when it can be filled or emptied.

There need be little difference in initial cost between a sweating kiln and a drum; but whereas rent and depreciation may be put at a maximum of 5 percent on a kiln, it will probably be wise to allow an additional 5 percent for depreciation on the machinery of the drum. Whilst the kiln will burn more fuel, and more expensive fuel, and take more labour, there will be a charge for the power necessary to work the drum. Again, the drum will occupy considerably less space than the kiln; a drum capable of dealing effectively with 50 quarters every fourteen hours is easily contained in a shed 15 by 35 feet, while a kiln capable of sweating 50 quarters every 24 hours would occupy some 800 square feet.

Comparison of Kilns and Drums for Sweating Barley— In a certain way the drum is more elastic than the kiln; supposing, for instance, that as much barley as possible has to be sweated in a short time in a 50-quarter drum. If a night man is employed, and the drum is started at three o'clock in the afternoon, the barley should be dry by 4 or 5 o'clock on the following morning; the night man can then start elevating his barley, which may take an hour and a half, and filling the drum, which may take a similar period. But, even under the most adverse conditions, a fresh lot will be started by 11 o'clock, and a third lot by 2 or 3 on the day following; this third lot should be dried by 3 P.M., so that between 3 P.M. on, say, Monday and 5 P.M. on Wednesday, 150 quarters of barley will have been sweated and elevated; although it must not be for gotten that barley sweated somewhat hurriedly at rather high temperatures and elevated warm, will need to be cooled before being stored, especially when the moisture content is not brought down very low.

The time occupied by filling and emptying the drum will, of course, depend upon the amount of hopper capacity beneath the drum, the capacity of the elevators, and the propinquity of the barley to the spout or spouts feeding the drum.

In the case of a kiln, however, it would be impossible for one man to turn and throw off 50 quarters of barley, so that it would be impossible to get more than one drying through in 24 hours without a considerable cost in extra labour, even if the kiln were capable of dealing effectually with its load in less than that time.

Capacity of Sweating Plant— The heaviest work of the sweating plant has to be done between October and the end of March. With early harvests a certain amount can be done in September, but barley sweated late in April will not be ready to steep before June, which is generally rather too warm a month in which to steep English barley, so that most of the barley sweated in April will have to be held over the summer for use in October. Taking seven months then as the maximum time available for sweating, a 50-quarter kiln, working six days a week, would dry some 10,000 quarters, and a drum of similar capacity, reckoning on five dryings every four days, some are 2,500 quarters. Of course, either of these quantities would be excessive for the requirements of a 10,000-quarter house, if considered in this way. As a matter of fact, however, barley is not likely to be bought evenly over the whole season, but rushed in when prices are at their lowest, so that in seasons when sweating is necessary, or where it is the rule to sweat all English barley, the 50-quarter capacity plant will not as a rule be excessive.

Expenses of Sweating Barley— Drying barley is not as a rule part of maltsters' proper work, and they will generally expect extra pay if required to load, turn, and throw off 50 quarters of barley; on the other hand, it will not pay, except in very large concerns, to keep men especially for barley work, so that it is generally best to arrange barley drying as piece work for the maltsters where a kiln is in use. With a well-arranged drum the amount of labour is so small that it may be included in the routine work.

It may be repeated here that some maltsters prefer to sweat all their English barley. For these the question is a simple one. Say 8000 quarters of English barley are to be made up. A 50-quarter kiln is erected at a cost of, say, £800; the interest on this at 5 percent will amount to £40 per annum, constituting a charge of 1.2*d.* per quarter sweated. Labour may cost i*d.* or 1.5*d.* per quarter, and fuel another 1.5*d.*, so that the cost of sweating, allowing for the extra power necessary for elevating, will work out at something not exceeding 5*d.* per quarter. If a drum is substituted for the kiln at the same initial expense, the capital charge for the same quantity, allowing for depreciation in plant, will amount to, say, 2.5*d.* per quarter; labour should not be charge able, and fuel will probably be reduced to something like 0.2*d.* per quarter, but the power required beyond that necessary for elevating may cost 1.5*d.* or 2*d.*, so that as regards actual cost per quarter there will not be much difference.

For those maltsters, however, who only have recourse to sweating in seasons when barley comes to hand with high moisture percentages, the cost question becomes more complicated, as in dry seasons no return will be got from the capital expended in sweating plant, which, however, is necessary in case of wet seasons. It is possible, of course, to make no provision for sweating, and trust to foreign chevalliers in wet seasons; but inasmuch as the price of the latter is always likely to be high in wet English seasons, this policy is often a costly one. Another way round the difficulty is to use the malt kilns for sweating barley; but this either leads to making the malt kilns unnecessarily large, or working the house below the capacity for which it was planned—both extravagant policies.

Reasons for Sweating Barley— As regards result, there is probably nothing to choose between barleys sweated in drum and on kiln, except in those cases where the sweating process is relied on as a means of maturation, when kilning is probably preferable, long sweating periods, low heats, and comparatively low velocities of the drying air all being conditions desirable for the maturing process.

Barley may be sweated with three objects in view:

1. For safety during storage.
2. To improve germination.
3. To improve modification.

1. *Storage*— If for any reason English barley is bought some months before it is to be steeped, it is necessary with really damp barleys—i.e., those with moisture percentages of from 17 to 20 (and sound barley may come to hand in wet seasons with moistures of over 20 percent), to bring the moisture down to some 12 per cent to ensure safety during storage. By safety I mean the avoidance of heating in bin or bag, however it is stored. With barleys of from 15 to 17 percent, it is safer to bring down the moisture by kiln drying if they are to be stored in bulk for any period of more than one month. Almost all English barleys are better for a sweat if they are to be "held over" from one season to another.

2. *Germination*— Damp barleys, with moistures of from 15 percent and upwards, often start growth on the floors with greater regularity if they have been sweated and stored for three weeks before being steeped. Corns which would not start growth until after three or four days on the floors, start with the rest after sweating and storage.

Under this heading also come early-harvested barleys, independent of their moisture percentage, and sometimes those which have been threshed out of the field and missed their natural sweat in the rick.

3. *Modification*— The part played by the sweating process in the maturation of barley is not yet thoroughly understood. There can be no doubt that the malt made from many barleys is improved in tenderness even to the extent of 2 or 3 lbs. of brewers' extract per quarter with the infusion system of mashing, if the barley has been subjected to sweating and storage before being steeped. With other barleys, it is very doubtful whether sweating effects any improvement.

Very roughly speaking, there are three classes into which unsweated barley may be divided:

(A) Those with mealy endosperms.
(B) Those with endosperms which appear steely, but become mealy after having been soaked· in water for two or three hours, and subsequently gently sweated till the moisture percentage is down to, say, 12 percent.
(C) Those with endosperms which are steely, and remain steely after the above treatment.

It is an open question whether any improvement in maturation, and, consequently, in modifiability on the growing floor can be made in the first of these classes. With the second class, there is practically no doubt that a great improvement is made, and with the third class it is equally certain that no improvement is made.

In practice, of course, deliveries of barley come in mixed in a great proportion of cases; that is, taking the arbitrary classes referred to, though some fine deliveries may consist entirely of Class "A," most average deliveries of fine barley will consist of a mixture of Class "A" and Class "B," while coarser barleys, such as it is customary to malt for running-beer malts, will consist of mixtures of Class "B" and Class "C." Class "C," of course, should be avoided altogether for malting.

Apart altogether from moisture percentage, then, it becomes questionable whether arty individual delivery of English barley should be sweated in order to effect an improvement in its modification on the floors. Inasmuch as 1 lb. of brewers' extract is normally worth about *4d.* to *4½d.*, and sweating should not cost much more than that amount per quarter, it would probably be economical to sweat all barleys, were it not for the decrease of weight, apart from loss of weight by expulsion of moisture which undoubtedly takes place under certain conditions during the sweating process. This subject is an important and an intricate one, and will be referred to again in connection with barley and malt, but it will suffice to say here that loss of dry matter does some times, though probably not always, occur when barley is sweated.

The loss of dry matter has been attributed to the splitting up of carbohydrate matter into carbonic acid gas and water under the conditions of heat and moisture obtaining during the sweating process, the amount of matter lost probably varying with the conditions of sweating, such as the amount of moisture present in the barley, the temperature and velocity of the air used to dry the barley, and the state of maturation in which the barley is prior to the process.

Taking the case, then, of a barley of Class "B," containing 15 to 17 per cent of moisture, a great improvement may be effected in the modifiability of the barley by sweating, accompanied by a certain loss of dry matter, the improvement in the quality of the resulting malt more than compensating for the loss in quantity. So that sweating such a barley will be the right policy, even though the condition of the barley may have been such that it would have germinated perfectly without having been dried.

On the other hand, if this barley had been up to the "A" standard, it would very possibly have been more economical to have steeped it without drying, as the loss of dry matter resulting from that process might not have been compensated for by any perceptible improvement in the finished malt.

With dry barleys, of moistures of 13 to 15 per cent, there is not so much loss of dry matter, but it is this knowledge that loss of weight may occur that restrains many maltsters from sweating their barleys unless they are forced to do so by a high water percentage, or unless they do so from the knowledge bought by long experience, that a sweat will be beneficial to certain types.

Chevallier foreigns are not, as a rule, imported unless they are dry enough to need no sweat; this is also the case with most brewing foreigns, but Yerli Smyrnas do come over in many seasons in large quantities in a condition that makes sweating desirable. Whether they can ever be bought at a price low enough to make them good value, when they contain sufficient moisture to make sweating necessary, is perhaps doubtful. It is probably unnecessary to say that if a barley has been heated, no amount of sweating, or anything else, will be of any use; but in the case of deliveries which come to hand containing a percentage of corns which have grown in the stack, sweating does improve the resulting malt. This improvement is made in two ways—firstly, because if the grown corns are steeped without having been sweated, they are likely to become a source of mould on the growing floors, and secondly, because, having been sweated, they often form fresh rootlets after having been steeped, and work up into fairly tender malt.

Storage of Barley between Sweating and Steeping— After sweating, it is generally customary to let barley rest in bulk for two or three weeks before steeping, as, if steeped sooner, there is danger that the start of germination will be very uneven.

The length of storage given, then, should never be less than three weeks unless it is absolutely imperative to steep the barley, and then it should only be done after very careful germination tests; beyond this necessary three weeks, the length of beneficial storage must depend upon the object with which the barley has been sweated.

For instance, if a damp, mellow barley has been sweated and long storage given, there is danger that the grain will work too quickly on the floors and an unduly high diastatic capacity with a large percentage of ready formed soluble carbohydrate matter will be features of the resultant malt, accompanied, probably, by high extract and high malting loss.

If, on the other hand, hard, dry white barleys are sweated with a view to maturation, the longer the storing period given, the better, up to, say, eight or nine months, after which there may be danger of impaired vitality.

Danger of Sweating and Storing in Bulk— If the sweating process has to be a somewhat hurried one, and some of the corns are not quite "sharp" to the bite after it is over, storage in bulk undoubtedly helps to average the moisture percentage through the heap, though care should be taken in these cases that the successive kiln loads are given time to cool down before being stored in bulk. If this is neglected the combination of dampness and warmth in the bulk may cause it to heat, and though such heating may proceed gradually for a long time without doing very much harm, and may very probably have the effect of increasing maturation, there is always the danger where barleys which are not quite thoroughly dried are stored in bulk while still warm, that the temperature of the bin may suddenly increase without warning, and a large percentage of the corns be killed; of course, should this occur a very heavy financial loss must ensue, the barley at once falling to grinding value.

CHAPTER V

STEEPING AND CONSTRUCTION OF FLOORS

IN a previous page some stress was laid on the necessity for keeping the steeping cisterns well above the working floors, and for keeping them all together, either on one floor or two, as may fit in best with the construction of that part of the malt-house. If this is done it will be possible, without going to any very great outlay either in initial expense or in maintenance, to provide attemperating plant and keep the steeping chamber up to a temperature sufficient to prevent delay in the colder months of the year.

Area of Steeping Cisterns— The area of the steeping cisterns should be rather more than sufficient to steep the quantity of barley for which the house is planned, as in very cold weather when pieces on the floors are thick and malt dries quickly on the kilns because draught is good it is very annoying to have the output of a house restricted by deficient steeping area. The driest barleys swell during steeping to the extent of quite 3 cubic feet per quarter of 448 lbs., and it is, therefore, wise to allow 14-15 cubic feet per quarter between the bottom of the overflow and the outlet of the cistern.

Practically all barley when steeped and well drained will run at an angle of 45°, but if the cisterns are hopper bottomed, as opposed to conical-bottomed, and the sides are put in at an angle of 45°, the corners, where there is likely to be most resistance, will of course be at a more obtuse angle, so that it is preferable to set the sides of the hoppers at an angle of 50° if the cisterns are to be really self-emptying.

Aeration in Steep— *Aeration* in steep has been somewhat fully discussed for the last ten years without very much having been done in this country. No doubt barley wants air in steep, and, if the steep is a long one and the steeping temperature is above 57° or 58° F. and the barley lies to a great depth in the cistern as must be more or less the case with hopper-bottomed cisterns, there is without doubt some danger of partial suffocation. The best way to meet these combined conditions in practice is to give somewhat frequent changes of water; with the deep cisterns, draining tends to drag the barley into a compact mass, and to cancel this the underlet should be used for rousing up.

Danger of Partial Suffocation of Barley in Deep Cisterns—Draining gives plenty of aeration as the receding water is followed by air which naturally penetrates to the interstices between the corns. Sparging through fine holes from some height above the barley also aerates, but it does not lighten the mass of grain in the same way and, therefore, should not be resorted to when there is any chance of suffocation in deep cisterns.

The whole subject of aeration was somewhat fully dealt with in a paper read by Messrs. JULIAN BAKER and W. D. DICK on "Some Observations on the Steeping of Malting Barley" (*Journal of the Institute of Brewing*, vol. xi. No. 5, June 1905), when the main points shown were that the water was absorbed in a much shorter time with aeration than without, and that the start of germination on the floor was distinctly quicker and more vigorous where aeration had been given. Whether, however, the quick absorption and rapid start does or does not shorten the flooring period necessary for the complete modification of any given barley, and whether the co-efficient of modification attainable with barley, which has been subjected to aeration during steeping is greater than, equal to, or less than that attainable with barley which has been steeped in the ordinary way, was not proved. Meanwhile most maltsters prefer a somewhat slow start, but the fact is that aeration has not been in practice either sufficiently long or sufficiently in general in this country to enable a very pronounced opinion on its merits to be given, although when stubborn barleys are being malted it is quite possible that aeration may be advantageous owing to the increased vigour with which the young pieces start germination.

Steeping Temperatures— The same thing applies to the composition of the water. Soft waters are somewhat more extractive than hard, and for that reason may be preferable. Tidal waters in which the proportion of chlorides is liable to variation may be dangerous, as a high chloride proportion may impair vitality. Where the supply is from a well, and a constant temperature of about 50°-54° F. is obtainable, the maltster has a considerable advantage over those who are dependent upon a town supply which may vary in temperature from over 60° F. in the warmer months to 40° F. in the colder.

There is no case on record of the vitality of barley having been impaired by low temperatures in the steep, but as the temperature falls below 50, the duration of the steep should be increased, as the colder the water the longer is the time necessary for absorption, other things being equal, and however complete the absorption, the start of germination is certain to be delayed when the steeping temperature has been a very low one. It is for this reason that it is so advisable to keep the cisterns together in one chamber which can be kept at a constant temperature, at least in the colder months.

High steeping temperatures of 60° F. and over, however, undoubtedly tend to impair the vitality of the grain, especially when combined with long steeping periods; and though no doubt the damage done depends upon a variety of conditions, such as the vitality of the barley, the thickness of the skin, and the length of steep given, yet it is extremely undesirable that the temperature of the barley in steep should ever rise to 60° F., although it is possible that the research of the future may show that this or even higher temperatures in steep combined with a short steeping period is permissible, and even desirable for certain barleys.

Effect of Temperature of Steep Water on Length of Steep necessary— In order to test the effect of short steeping periods on high temperatures, an experiment was recently made on a kiln-dried Chevallier barley of the 1906 crop. The approximate moisture percentage of the barley was 10, and 99 percent of the corns grew in the ordinary Coldewe germinator. Equal quantities of about 150 grammes were steeped at a constant temperature of 67°-68° F. for periods of

17, 27, 37, and 47 hours, the water being changed every ten hours, except during the last seventeen, when no change was given.
The moisture percentage after steeping was as follows:

		Moisture percentage after steep	
A	47 hours steep	45 percent	Water changed at 10, 20, and 30 hours
B	37 hours steep	41 percent	Ditto ditto
C	27 hours steep	40 percent	Ditto at 10 and 20 hours
D	17 hours steep	37 percent	Unchanged

After careful germination in the Blaber cupboard for four days, at temperatures varying from 50°-60° F., but constant for the four cylinders, the percentages of dead and slow starting corns were as follows:

	Dead	Slow Start
A	10 percent	2 percent
B	7 percent	2 percent
C	5 percent	2 percent
D	3 percent	2 percent

and though the D sample was then very dry, it could probably have been nursed up by skillful flooring and sprinkling to fairly tender malt.

Another experiment carried out on equal quantities of the same barley, but at a temperature of about 45° F. showed the following results:

		Moisture percentage after steep
A	96 hours steep	42
B	72 hours steep	40
C	48 hours steep	38
D	24 hours steep	33

and in this case, though of course D was very under-steeped and could not have been successfully germinated, no corns were killed in any of the lots.

It is not desired to give too much importance to these very simple experiments, and they are only quoted as bearing out the remarks made above on curtailing the steeping period in proportion as the temperature of the steep rises above 58° F.

Boiling in steep often occurs when the temperature is too high, and is, of course, caused by the liberation of carbonic acid gas resulting from the splitting up of carbohydrate material. When it occurs, the steeping period should be shortened, and as much aeration as possible given; but it is invariably a sign of high steep temperatures, and its occurrence should be taken as a warning that colder water must be employed.

Addition of Lime to Steep Water— Freshly harvested English barley is coated with a soft mucous layer, which during sweating, handling, and storage gets to a great extent rubbed off. When, however, freshly harvested barley has to be steeped, this waxy layer retards absorption of water to a certain extent; the addition of lime to the steep water helps to remedy this.

The best way to add lime is to arrange a tank commanding the cisterns, the tank being constructed so that ordinary quicklime can be held in the bottom. Water is run in and allowed to become saturated and then run off into the cisterns after it has become bright, the vent pipe being placed at a sufficient distance from the bottom of the lime tank to allow all the lime to settle.

The practice of adding lime water to the steep is not a very general one, as the additional plant, though small, is often difficult to fit into existing buildings; when a high percentage of lime water is used in the steep, a certain amount of lime is always present in the grain on the floors. This does not matter much on the floors except during withering when the old piece is allowed to become very dry, but as the malt is dried on the kilns the dry lime becomes a great nuisance while turning is going on, especially during curing. Where kiln turners are used this does not cause much inconvenience, but even then throwing the malt off the kilns becomes a very unpleasant business. Liming in steep should not be confused with the use of the chloride of lime as a disinfectant on the floors, the latter process being a dangerous one, which may injure the vitality of the gram.

Duration of Steep and Draining— It is difficult to give any rule for determining when barley has been in steep long enough. The corns when firmly pressed by the finger and thumb should open up without too much resistance, and it should be possible to make the young rootlet exude by a gentle pressure of the thumbnail on the embryo. Gentle pressure and firm pressure however are terms which are greatly qualified by the personal equation, and experience is the greatest factor in the determination. It is important that the steeped barley should be well drained before being discharged on to the working floors, and as long an interval as possible should be given between running off the water and emptying, an overnight drain being of considerable help when the temperature makes it possible.

The following approximate steeping periods may be taken as roughly representing those necessary for the respective kinds of barley, given a temperature of from 50-56° F. and no aeration beyond that which can be given in any well-constructed cistern furnished with a sparge arm and underlet. Newly-harvested English, 48-60 hours; kiln-dried English, 48-60 hours, unless it has been in store for over six months, when more than 48 hours will seldom be advisable. European Chevalliers, such as Hungarians, Saale, Moravian, and French, the same as English, unless perceptibly drier, when rather longer may be advisable.

Chevallier Chilians, Chevallier Californians, and New Zealands, 60-72 hours.

Brewing Foreigns of the six-rowed varieties, such as Brewing Californians and Oregons, Yerli Smyrna, Indians, Brewing Chilians, Spanish, North African Tripoli, and Ben Ghazi, 60-72 hours. The coarser types of six-rowed, such as Algerian and Tunisian, and the two-rowed brewing types, such as Ouchacs, Anatolians, and Syrian Tripolis, generally require 72 hours, unless the season has been specially suited to them, and they sometimes benefit by as much as 90 hours.

There is a very good rule, however, that it is an exception when the treatment required for the barley produced in any country requires the same treatment in two consecutive seasons either in the steep or on the floors.

Screening— It is advisable that the last process before steeping barley should be screening it. All barley contains a good deal of dust, and barley which has been kiln-dried and stored, a great deal. Practically this fits in well with malt-house routine, as it is much more convenient to screen barley immediately before steeping than when it first comes into the malt-house. If for any reason, however, it is impossible to screen before steeping, the water should be run up to the required level first and the barley run into it slowly, so that as much dust, chaff, and light barley as possible may be floated out.

Swimming in— This "swimming in," as it is called, is rather a long process and should not be done unnecessarily, but if it is done, it should be thoroughly done, and it is as well when it is over to open the outlet valves a little and wash through from the sparge arms for twenty minutes or half an hour, in order that as much dirt and foreign matter as possible may be washed away.

Every time the cisterns are emptied the draining plates which constitute the false bottom should be removed and well scrubbed, as they are always coated with slime after a steep. The same rule applies to the perforated plates above the overflows, where these are used, and the cisterns themselves should be well washed down when empty and then rinsed out from the sparge.

Growing Floors— Before starting to write about flooring it will be as well to make a few remarks about the floors themselves. In old buildings they were very low, but in the more recently constructed houses 7 feet is generally allowed from floor to floor, thus giving 6 feet of clear head room, and allowing 1 foot for girders and floor. The surface is almost invariably either cement on concrete, or tiles, and though the latter are a good deal more expensive there can be no doubt that they are more satisfactory if well laid than the best laid concrete. Whether cement or tile floors are used it is absolutely essential that they should be dead level, as otherwise the water will find its way to the low parts after sprinkling, with the result that growth will be uneven. If tiles are used they should be set diagonally and not square, as in the latter case any slight inequality will tend to catch and break the edge of the turning shovel. It is also of importance, if possible, to arrange that each cistern should empty its load on to one floor. If the load has to be divided there is always the chance that one floor will get more than its proper share. It is as well that the bricks in the walls of the working floors shall be glazed for a distance of 9 to I2 inches from the floor, so that the part of the wall in contact with the growing grain should afford no chances for mould to collect, and also that it may be easily washed down.

It is impossible to insist too strongly upon the desirability of keeping the growing floors free from obstructions. Pillars of some sort there must be to carry the transverse girders, but let there be as little else besides as possible. Stairs, for instance, at the two ends of the floors are a great nuisance, yet, as far as we know, only one maltster is wise enough to carry them up outside the building, though the increased cost need not be a very great matter. Loading elevators again take up a great deal of room, and afford plenty of scope for untidiness unless the man with the broom is extremely conscientious, but here again we have architects planting them down and cutting up the floors for them, though it would be often quite a simple matter to take them up through the kilns. If inside stairs are necessary, a cat-ladder in one corner should do all that is wanted, though in many maltings two flights are put in, one at each end, which take up many feet of valuable space and need incessant sweeping, to say nothing of the amount of green malt crushed and spoilt on them. Water-pipes must be carried along each floor for sprinkling purposes; the taps should be about 2 feet 6 inches to 3 feet from the floor, and beneath each tap a cast-iron perforated draining plate of about 1 square foot should be set on the floor, communicating with a waste-pipe. These plates should be taken out and well scrubbed about once a week. This is essentially a case in which the simplest arrangement is the best, and all obstructions should be avoided.

Windows— The windows form an important part of the construction in every malt-house. In small houses, where it is sometimes difficult in winter to keep the temperature of the working floors high enough, double windows with an air space between are good, but in large houses this difficulty is rarely met with; for these, two types of windows are superior to others, one being twin glazed frames swinging together on central pivots, capable of being opened to any extent, and held in position by a thumbscrew on a sliding frame; the other, plain boards with squares of glass in the centre arranged to slide in wooden grooves. The point of the first type is that when there is any wind, the windows can be set at an angle sufficient to break its force from the growing floors, while when there is no wind and it is desirable to give all the ventilation possible, they can be opened to their fullest extent. This does not apply to the sliding shutters, but these allow of a canvas blind being fastened over them; such blinds, if well arranged, break the wind even more effectively than the windows themselves, and have the advantage that they can be moistened with a syringe in warm weather so that the air passing through them is cooled by the evaporation which takes place. These blinds are very effective and very cheap, the bags used for importing brewing Californian barley serving admirably for the purpose, but of course it is impossible to use them where windows open inwards. On the whole it is probable that the sliding type of windows with blinds are preferable to the pivot windows, and they are simpler in construction and somewhat cheaper to install.

Coloured glass, especially blue glass, although yellow is far less actinic, has been recommended for malt-house windows, and it is a fact that strong sunlight is undesirable on the germinating grain; if there are curtains, however, this will be easily preventable, and after all one object of windows is to admit light.

To sum up, the chief points about floors are that they should be level, and, if possible formed of tiles laid diagonally; that the part of the walls with which the grain is in contact should present to it a smooth surface, and one capable of being quickly and easily cleaned; that one floor should be reserved for each steep; that they should be absolutely shut off from malt dust and barley dust, and as far as possible clear of all obstruction, such as staircases and elevator shafts. Thorough ventilation is of course necessary, and for this reason windows should be plentiful, and the breadth of the floors should not be much more than half their length. Cellar floors are not advisable in badly draining soils or in damp places, but otherwise, provided effective ventilation can be maintained, they are rather desirable than otherwise. It should be stated here that no slight is intended upon small houses by the constant allusion to large ones, which is made because, other things being equal, it is more difficult both to design and to work a large house than a small one, though the latter gives the greater scope for economy both in construction and in subsequent working. It would be absurd to suggest that the size of the house can have any effect upon the quality of the malt turned out.

CHAPTER VI

FLOORING

It must be admitted that both the barleys which may be worked and the atmospheric conditions existing at any given time are so variable that to give any definite rules for the flooring process must be to get on to dangerous ground, and it is proposed here, first, briefly to consider the main aims with which flooring is carried out; secondly, to divide it into the stages wherein the most critical times appear; and thirdly, to hint at the treatment most likely to be conducive to satisfactory results from the time the steeped barley is emptied on to the growing floor to the time when it is loaded on to the kiln. It will be necessary before doing this to make a brief allusion to the physiology of the germinating process in the interval between emptying the cisterns and loading the kiln, in order that the reasons for certain processes may be clearly understood.

Enzymes— During flooring certain enzymes are secreted, which, provided by nature for the development of the growing plant, are also necessary for the malting and brewing processes, and it is the regulation of these enzymes to the proportions necessary for malting and brewing that must form the chief object of the maltster.

The best known of these are *diastase*, whose function is to act upon the malt starch in the mash tun forming it into more or less easily fermentable types of sugar; the *proteolytic enzyme* or, enzyme whose function it is to act upon the nitrogenous constituents of the barley, forming from them those bodies necessary for yeast nutriment and possibly also the "head " and palate fullness in beer; and *cytase* whose sphere of utility is confined to the malting floors and whose function is to dissolve the walls of cellulose surrounding the starch cells of the endosperm of the barley corn, and thereby set them free for subsequent saccharification in the mash tun.

Whilst a vigorous activity on the part of these enzymes has to be fostered, it is not desirable that their secretion should be either too rapid or unduly prolonged, and as there is little doubt that they are up to a certain point all affected by the same conditions, it must be the maltster's object to ensure as far as possible those which will conduce to their healthy development, without allowing them to get beyond his control. If they do go too far the results in the finished malt will be undue formation of (1) the carbohydrate matter provided for the sustenance of the young plant, accompanied by waste growth of rootlet and acrospire at the expense of the endosperm starch, which is potential brewers' extract; (2) diastatic capacity, unless this is very carefully regulated on the kiln, and (3) peptatic activity resulting in a higher percentage of soluble albuminoids than is considered advisable for malt destined for beers brewed on the English system.

Malts in which the enzymic activity has been too great upon the floors, are distinguished by the high percentage of their solids soluble in cold water, and as many chemists express these as ready-formed soluble carbohydrates a certain amount of misapprehension has grown up in the minds of maltsters on this subject; the matters soluble in cold water include ready-formed soluble carbohydrates, soluble uncoagulable albuminoids, ash, and acid, and the fact that the percentage of all these is high is a criterion that enzymic activity on the floors has been excessive. Some barleys naturally form greater percentages of them than others, but what may be regarded as certain is that given a constant barley, which can be satisfactorily malted with a given percentage of cold water soluble solids, any substantial rise in such solids in the resultant malt will be attributable to uncontrolled enzymic action during flooring or in the earlier stages of kilning *unless such barley has matured between the malting of one lot and the other.* Such maturation does of course take place most seasons between October and March, and it might be unfair to condemn the flooring of a malt made up in March on account of percentages of ready-formed soluble carbohydrates and soluble uncoagulable albuminoids higher than those formed by malting the same barley in November.

"Forcing" is the word generally used to indicate *insufficient* control of enzymic action on the floors, and a good deal of ink has been used to teach the maltster at what temperature the grain should be kept, how much, or how little sprinkle should be given to it, and how long it should be kept out on the floors; but let it be written at once that there can be no one rule for the flooring process any more than there can be one universal type of barley of constant size, quality, weight, and maturation, and that what might constitute a forcing treatment for one type would as likely as not be quite the reverse for another.

The object of flooring malt, practically considered, is to secure a dry mealy condition of the endoperm starch with as little loss of weight by rootlet and acrospire growth and evolution of carbonic acid gas as possible. The shorter the time in which this can be accomplished the better both for the malt and for the maltster. Some barleys are capable of attaining satisfactory modification in from nine and a half to ten clear days from the time of emptying the cistern. Others require twelve, others fourteen, though the necessity for the latter period is generally caused by loss of time from working too cold on the floors after emptying from the cistern in very cold weather.

It must be said, judging from a certain amount of experience in the flooring of barley from many different countries, that a temperature of 70° F. in the growing grain is never advisable, though sometimes unavoidable, and that temperatures of between 65° F. to 68° F. are never welcome except in certain conditions of the young piece which will be alluded to later. Equally certain is it that all temperatures below 50° F. are too low, and that a great deal of unnecessary knocking about is often given from a mistaken idea that "turning keeps the sugars down," and that a good deal of grain is often kept out for 12 and 13 days when it could be made into equally good malt and undergo less malting loss and liability to mould if it were loaded to the kiln two days earlier, though of course this latter method is far preferable to the other extreme of high flooring temperatures and short periods which is still in use in some, though now probably in only very few, places.

Implements— The implements used for working malt on the floors vary somewhat in different parts of the country, but the following list comprises the most useful of them:

1. The turning shovel, for use on all occasions when plenty of mixing and aeration are desirable.
2. The loading shovel, similar to but rather larger and heavier than the turning shovel, and used for lifting the grain into the barrows, carts, or baskets used for transferring it from one part of the floor to another. The usual width of a turning shovel is 14, and of a loading shovel, 16 inches.
3. The tin shovel, whose use is restricted to breaking out the couch and chopping over the grain before rootlets are formed.
4. The fork, for use only during the last day or two out when the grain is comparatively dry and there is no moisture on the floors.
5. The stick-plough; this may be either as its name implies simply a stick to be used for lightening up the grain and moving the bottom layer from the floor, when a turn would not be advisable, or a miniature turning shovel with a blade only 2 or 3 inches broad, which without giving as much aeration as a turning shovel moves the grain more than the stick, and is used chiefly during the withering stage.
6. The rake, or pull-plough, which is an iron plough to be dragged between the floor and the bottom layer of grain, and is useful for lightening up a floor or before turning after the malt has been lying untouched for some time and more or less felting of rootlets may have occurred on the floor.

For transferring the growing corn from one part of the floor to another either barrows, carts, or baskets are used. Barrows should have very broad wheels and india rubber tyres, and should never be wheeled except on 2 or 3 inches of malt. Carrying-carts entail more labour but some maltsters think there is less danger of crushed corns if they are used. For baskets, rails have to be suspended from the top of each floor, to which they may be hung from grooved rollers; they are useful on very narrow floors.

Each lot of growing grain is technically called a "piece," thus the young pieces, the watered pieces, and the old pieces are constantly referred to during malting operations, and a thermometer should be kept in each piece from the time of emptying from the cistern to the time it is loaded into the kiln.

The flooring process may be roughly divided into three parts, start of germination, sprinkling, and withering. The divisions are somewhat arbitrary; some barleys can be satisfactorily malted without the addition of any sprinkle on the floors, and while some maltsters are inclined to make a regular fetish of the so-called withering process, other (and equally successful) men practically disregard it altogether, and in point of fact rather pride themselves on loading this malt to the kiln "fresh."

Moisture Percentage in Green Malt— The percentage of moisture present in the grain immediately on being emptied from steep varies in practice from about 42 to 44 percent, and it is interesting to note that while few maltsters load with a lower moisture percentage than 40, some reckon on never greatly exceeding this figure, while others constantly work for a percentage of 42 to 43, so that it is obvious that withering to comparative dryness is by no means universal.

At present only cold weather malting will be discussed, for under adverse conditions of temperature special precautions have to be taken, and the whole process has to be considerably altered.

Couching— Now assuming that the steeped barley has been drained at five o'clock, and is to be emptied at ten o'clock in the morning, two methods of dealing with it at once present themselves; either a couch 2 or 3 feet high may be formed, or the piece may be spread out on the floor at a thickness of, say, from 12 to 15 inches.

It is possible to maintain an air temperature of about 50° F. in large maltings, even in very cold weather, by careful control of the windows owing to the heat evolved by the growing grain. In small and exposed maltings the temperature sometimes falls very low in cold weather, and in these cases of course the start of germination is delayed, but they need not be considered here.

If, by regulating the temperature of the steeping chamber, the latter has been kept up to about 54° F., the grain will empty at very little below this temperature, but in cases where the steep water has been very cold it will generally be best to allow the grain to be in thick couches to accumulate a growing temperature. But even then the thick couch system constitutes the lesser of two evils, for where resort is made to it the grain in the middle of the heap is likely to accumulate a good growing heat, while the corn on the outside remains cooler and is becoming dry by evaporation, and until the couch is broken out the inside corn gets little or none of the aeration which is very necessary at this stage. If, on the other hand, the grain is spread out comparatively thin, say, 12 to 15 inches, immediately on leaving the cistern, the temperature will be more constant, and though the start of germination may be slower it will probably be more even, unless the couch has been broken at exactly the right moment.

Necessity for Aeration of Young Pieces— Even when the pieces are lying at low temperatures, 50° to 52° F., they generally benefit by being either "chopped over" with a shovel or furrowed with a stick plough once or twice a day to ensure aeration and mixing.

Whether the piece has been couched thick and broken out, or couched thin arid chopped over at intervals, it will want very careful watching when the temperature has risen to 56° F. or 57° F. and the corns have started "chitting," i.e., when the young plant is just discernible at the embryo end; for it is at his stage that the increase in vitality may be either gradual and even, or very sudden, or sluggish according to the character of the barley and the treatment given.

Sweat on Growing-Pieces— If the growth progresses evenly no trouble will be experienced, two turns a day—say at 6 A.M. and 4 P.M.—will generally be advisable, while the rake may be used occasionally if necessary according to the amount of rootlet formed. This is the easy side of flooring, and given good malting weather, it may generally be ensured by buying barleys of fine quality and even in size, condition, and maturity. But a rather different state of things must be the rule in most houses. It is delightfully easy to malt these fine barleys, but the skill of the maltster comes into play when he is called upon to make good malt out of those barleys which do not lend themselves quite so easily to the process. In these cases the barley may work in very much the same manner at first, till after having been out for some 48 hours it has started growth evenly and been turned three or four times. Supposing then it has been duly turned on the evening of the second day at a temperature of 57° F., the next morning it is lying perhaps at 61° F. with a certain amount of "sweat" showing in the corns, proving that there has been work done, accompanied by formation of carbonic acid gas and water during the night; then some two or three hours after the morning turn there may be a quick increase of temperature with a great deal more sweat and rapid rootlet formation. If this rootlet is thick and bushy it may benefit by being well turned, thinned out, and raked; but while this drastic treatment may be the making of some malts it is the ruin of others, whose lower vitality or enzymic activity—possibly synonymous terms—is not sufficiently strong to allow them to stand it and in these cases the temperature falls, respiration ceases, and the rootlets tend to wither, so that the piece has to be thickened up again and left untouched for some time, until it shows signs of recovering its activity.

Too Vigorous Growth in Young Pieces— This "rushing," as it is sometimes called, constitutes one of the most critical periods of the grain on the floors. It may occur any time after the piece has been out 24 hours or it may not occur at all; it may be kept up for 48 or even for 60 hours, or the sweat may die down and the rootlets curl up and wither within 24 hours of its start. If it is sustained it can be dealt with by constant turning, furrowing, and raking, and the barley is very likely to come through it with a strong bushy rootlet and in good trim for sprinkling when about five days old. But in lower vitality barleys there is great danger that it may be checked too soon, with the result that the piece has to be nursed back to life, as described above, and valuable time thereby wasted. Some pieces can not be nursed back to vigour, when hard malt must be the result.

At the risk of undue reiteration it must be emphasised that barleys behave very differently at this stage of the flooring period. Some work evenly, some rush stupendously and fall off without warning, others rush stupendously and fall off gradually, while others again rush moderately; it is with these last two classes, and especially with the last that high temperatures, viz., 65°, to 68°, or even 70° F. may be permissible and even beneficial to the resultant malt. For if they have been judiciously treated during the strong germinative period, and sprinkling is postponed until they have got it over and are just commencing to wither, then if the sprinkle liquor is given in the right proportion there should be little trouble in bringing the piece to full modification without much difficulty.

Forcing the Young Pieces— Of course if the young pieces are allowed to lie unturned when they are sweating vigorously the rush will be forced on. This is simply bad flooring, though in barley of low enzymic activity it may sometimes be advisable as the lesser of two evils, but as a rule barleys which rush and on being vigorously treated fall off suddenly are not suitable for manufacture into malts for any class of pale or running beers, and are generally characterised by low enzymic activity. Chevallier Californians are often a case in point, for they start germination with tremendous activity, are very easily checked, and almost invariably produce malts characterised by low diastatic capacity and a low percentage of matters soluble in cold water.

Rise in Enzymic Activity Coincident with Vigorous Growth— In enumerating the enzymes whose action is developed on the floors no mention was made of oxidase which may be regarded simply as an oxydising agent by whose aid some portion of the starch is enabled to combine with the oxygen whereby it is degraded into carbonic acid gas and water.

$$C_{12}H_{20}0_{10} + 12\ 0_2 = 12\ C0_2 + 10\ OH_2$$

It is obvious that this reaction is dependent upon the capacity of the starch to combine with oxygen, which would be impossible without the assistance of oxidase. The heat evolved on the growing floors, and the rushing of the young pieces which occurs in consequence of too vigorous action is no doubt due to this agency, and, although as far as is known no estimations of the increase in the energy of oxidase during flooring have been carried out, yet, inasmuch as the environment conducive to the activity of one enzyme is most probably equally conducive to that of the others, attention may be called to a paper read by Mr. GORDON SALAMON on "Some Experiments in Malt Making" (*Journal of the Institute of Brewing*, vol. viii. No. 1,

January 1902), where the rise in the diastatic activity noted in green malts between the third and fourth days on the floors was, in some cases, very pronounced.

Sprinkling— Whether there has been a marked "rush" in the young pieces or germination has proceeded with regularity, by the time the grain has been on the floors for about five days there is usually a falling off in vitality consequent on the evaporation which has been proceeding from the time the steeped barley was emptied from the cistern. The evaporation takes place with more or less rapidity according to the relative drying power of the atmosphere and other circumstances, and is chiefly noticeable in the rootlets which, especially after a rush, have a very yellow and draggled appearance. But it may also be realised by breaking the corns open with the fingernail when the starchy contents of the endosperm will feel distinctly drier than they did when the moisture content was higher. Actual moisture determinations carried out in the laboratory are not of very much value to determine the loss of moisture effected by evaporation during this early stage as, aside from the liability of green corn to lose weight apart from moisture during the determination, there is a varying amount of moisture constantly present on the surface of the corns due to the sweat which is an accompaniment of the process of modification proceeding in the malt-house floor. Experience must remain the greatest factor in determining exactly when it may be necessary to commence sprinkling, but it may be safely written that the rule is that it should never be done until the pieces are comparatively dry and, preferably, should never be done when the temperature recorded on a thermometer resting in the growing grain is above 60° F.; the exceptions are barleys which tend to fall off very suddenly after a short period of very quick growth, these sometimes being benefited by gentle treatment on the floors and constant small doses of sprinkle liquor. Such barleys, however, are to be avoided for malting, and with normal malting barley and normal weather five days will generally have expired from the time the grain is emptied to the time when use of the sprinkling can becomes necessary.

The quantity of liquor which it is advisable to give any barley must vary very greatly according to the barley, the weather, the malting process, and, to a certain extent, the length of steep given. It is not intended to enter into the question of steeping short and making up by earlier sprinkling and larger quantities of sprinkling liquor, because in very nearly all cases it is most probable that steeping should be carried out on the premise that a certain duration is most suited to each individual type of barley, but it may be said here that with barleys of low vitality it is often preferable to curtail the steeping period and make up with extra sprinkling, while with very hard dry barleys, of which perhaps Ouchacs and Anatolians are the best examples, it will be found to answer better to give a very long steep and proportionately less water on the floors. Again, though it is not advisable further to complicate what must inevitably be a very difficult subject by alluding to changes of temperature and hygrometry, it is obvious that in spring, for instance, when the weather is warm, all windows are open, and the air has a relatively high drying power, more sprinkle will probably be necessary than in winter when all windows are shut, or only opened during turning, and the air in the malting floors is nearly saturated with moisture. Again, with regard to the malting process, if it is thought advisable that the flooring period should continue for fourteen days, and the pieces are allowed to lie thick with only one turn a day, it will be absurd to give as much sprinkle as if a shorter flooring period of, say, ten to eleven days with constant turning and relatively thin pieces were contemplated. If the former process is necessary for the satisfactory modification of the barley, and growth is even, it may be as well to give the water in two, three, or four lots, say, on the mornings of the sixth, seventh, eighth, and ninth days in small quantities, in order that the pieces may be kept fresh until withering is to be allowed to commence on, say, the twelfth day out. But if rootlet growth be uneven, and the water is given in small doses, there is danger that those corns with the greatest rootlet development will get more than their share of water at the expense of the corns less developed in this and other respects, and which really require more than the former to stimulate a somewhat sluggish development. If, however, satisfactory modification can be secured with a flooring period of ten or eleven days, there is no question that it is very much better to get sprinkling over as soon as possible, and even then if the kilns are to be loaded ten days from emptying, thin floors and constant turning will be necessary to prevent unduly high flooring temperatures. Under these conditions it will be best to give the pieces at least half the destined amount on the morning of the sixth day out, and if this quantity is larger, it will be best to divide it into two portions, and after giving the first have the piece thoroughly well turned, when the second half may be given, followed by another good turn. If the quantity thus added be great, and there are cases where barley has benefited by receiving as much as six gallons to the quarter in this way, though they are the exception, it is very necessary that about two hours after the second turn the floors should be carefully examined to see whether the water has been absorbed. If it has, and the surface of the floor is only moist, it is generally as well to draw the rake through the watered pieces; but if it is wet, it is a sign that the amount of water was excessive, and another turn should be given, and the amount reduced in the next piece.

Heavy Sprinkling— With these large additions of sprinkling liquor, involving as they do much turning to ensure thorough and regular absorption, some hours of quiet are generally needed before the recommencement of active germination. If the amount of water added has been excessive, it will be necessary to keep the pieces lying thin and to turn them fairly constantly for the day or two following sprinkling, but if it is thought necessary to add more liquor this should be done as soon as possible, preferably on the morning following the first sprinkling. The number of the day is at first somewhat confusing, unless it is remembered that the day on which a piece empties from the cistern, which is sometimes alluded to as the first day, it is no days old—consequently on the second morning it is one day old, and if watering be carried out when the piece is five days old, it will be on the sixth morning. If then a piece is to be loaded to kiln ten days old, loading will take place on the eleventh morning, and when this is to be done no water should be added after the eighth morning, except on those unfortunate cases where the amount of sprinkle liquor necessary has been under estimated, and there is a choice of evils between under modification and the chance of loading to kiln too damp. Theoretically, of course, this could be avoided by postponing loading till a day later, but practically there must be more or less routine in every malt-house, and keeping a piece on the floors for a day more than was calculated upon may make them unpleasantly full for that period. If a piece has to be kept out in order to get rid of its superfluous moisture it will be necessary to keep it fairly thin, and this will possibly necessitate keeping the floors following it unduly thick, unless either the weather is very accommodating or the floors are being worked considerably below their capacity.

As a matter of fact, at the commencement of a season, or with the first lot of a parcel of foreign barley, the amount of sprinkle liquor necessary must be more or less of a problem, although with subsequent lots of the same barley it should be easy to estimate; and it is for this reason that it is so advisable to malt as much of one type of barley as possible in one malt-house, or at any rate, if it is necessary to malt different sorts to keep one cistern floor and kiln for each for as many consecutive steepings as possible.

Duration of Sprinkling— It will be evident from the above that the duration of sprinkling must depend upon the length of flooring to be given. If a long flooring period is to be given it may be preferable to distribute it in small quantities during three or four days, and it will be easier to correct any superfluous or deficient quantity given the first day than if all the liquor has to be added during a maximum period of three days, and preferably one of 48 hours. Inasmuch as it is impossible to get some barleys satisfactorily modified in a flooring period of less than twelve or thirteen days, sprinkling may be an easy matter, but when barleys are to be malted which can be well modified in ten days sprinkling must be a most important process, and one in which the skill and experience of the maltster must be brought into play.

Duration of Flooring— It may be that some barleys are capable of yielding equally satisfactory malt, whether made on the long or the short flooring system. To keep these out for the long period would be waste of time and weight in the resulting malt; it would be still worse to try to malt on the short flooring system a barley which was incapable of satisfactory modification on the time given. "Forcing" is the term generally applied to trying to do this, and there is no doubt that on the floor system of English malting most forcing is done by maltsters giving their pieces more sprinkling liquor than they can make use of, with the result that high flooring temperatures ensue, enzymic activity is great, and acrospires grow out during the days between sprinkling and loading. This is especially the case when the excessive sprinkle liquor given is helped by scanty turning and thick pieces during the days preceding loading the kiln, and though the rootlets formed under these conditions may not be, and frequently are not excessive, undue growth of acrospire almost invariably results. If kindly, well-matured barley is treated in this way the result is both loss of weight and forced malt, but even if it is used as a means of making coarse barley tender, the desired effect will not be gained in nine cases out of ten by the forcing treatment, as the pieces lying at considerable depth without turning will modify unevenly, and there is always the chance that while many corns show excessive development of acrospire, many others will simply be suffocated for want of air in the atmosphere of carbonic acid gas evolved by the piece.

Sprinkling Methods— With regard to the actual process of applying the water to the pieces, two systems are in vogue. One is to have hose-pipes fitted to the water taps on the floors, to which ordinary roses are fixed; the pipes are of sufficient length to command the whole portion of the floor usually occupied by the pieces during the period of sprinkling, so that the process resembles that of a gardener watering a bed of flowers. The only way to estimate the amount of water given is to have meters fixed at each pipe. There is not much to be said about this method except that it is not so good as the more usual one of sprinkling from ordinary large watering-pots, capable of holding about four gallons when full. These watering-pots are furnished either with roses or with long perforated pipes, the latter being preferable if the perforations are not too large, as they admit of more even distribution of small quantities of water.

Spraying to counteract effects of Evaporation— Sprinkling may be done for either of two reasons, the first of which has been discussed above, and is, briefly, to increase the moisture content of the whole piece, revive the rootlets, and stimulate enzymic action. When these are the objects, it has been seen that comparatively large quantities of water are given, and sprinkling is immediately followed by turning, so that the water may be well mixed into the growing grain. The second reason is of much less frequent occurrence, and is simply to bring the top layer of the grain to a moisture content equal to that of the rest of the piece which it protects from the causes of evaporation. If, for instance, a piece has been turned at six o'clock in the morning and is examined at four o'clock in the afternoon of the same day, especially when the windows are open, it will generally be noticed that whereas the bulk of the grain is fresh, the very top layer is dry, and this state of things is generally remedied by mixing the whole piece together by turning. Sometimes, however, it would obviously be better not to turn, and then it may be advisable to spray the surface with just as much water as will moisten it without penetrating below the dried top layer, and with this object in view it is as well to have in every malting two or three 'pipes with very small perforations, so that the distribution of the small quantity of water necessary over a large surface may be as even as possible. The objection to the spraying of course is that the sprinkler must leave footsteps behind him on the piece, which will not be obliterated by subsequent turning or furrowing, but it is wonderful how little trace a skilled maltster will leave on a floor after walking over it and moistening the surface, and if the work is carried out by a man who understands it, the amount of corn left lying solid will be very small.

The Last Five Days. Varying Flooring Periods Suitable for different Barleys— In the preceding pages it was pointed out that the length of flooring to be given to any barley must, as a rule, be a considerable factor both in the amount of sprinkle liquor necessary and in the duration of the sprinkling process. It is very difficult to tell when sprinkling will be over except when several successive wettings of the same barley are being worked during a period of practically constant climatic conditions, when the treatment of the first piece which gives really satisfactory malt both by analytical and physical tests, may be taken as a sort of pattern for the rest while those conditions are maintained. In writing about these last days on the floors, therefore, it will be as well to take the cases of barleys requiring long and short periods of flooring to ensure satisfactory modification.

The Short Flooring Period— To begin with the latter class. A barley is emptied from the cistern on a Monday morning and is watered at five days old on the following Saturday morning (the sixth morning out). More water is given on Sunday morning, the seventh, and again more on Monday morning, the eighth. On being examined on the Monday afternoon it is thought that it has had sufficient sprinkle, the temperature is 63- 65° F., the rootlets fresh and vigorous, and when individual corns are broken open lengthways, the contents of the endosperm are found to be soft, damp, and very fairly mealy in the centre, though in most corns the starch in the ends far removed from the rootlets is hard, and inclined to roll up into a ball instead of mealing down when rubbed between the finger and thumb. If the corns are cut lengthways it will be seen that where the starch feels "mealy" and rubs down flowery, it looks white, but where it feels hard and rubs down doughy it looks translucent, and in very many corns this translucent appearance is not only evident at the ends, but also all round the inside of the husk except just opposite the scutellum, where the parasitic embryo joins the endosperm. Now if the piece has been sprinkled and turned at 6 A.M. and raked at 10 A.M., and there is no marked increase of temperature or growth of rootlet, it will probably be sufficient to give it a good turn at 4 P.M. If, however, there is perceptible growth of rootlet since morning and the piece looks at all rough, it should be furrowed before being turned, and if the temperature has risen more than, say, three degrees since the morning turn, it will probably be advisable to thin it out somewhat. This can be done by skillful turners by starting a long turn forward and gradually shortening the distance the grain is thrown forward from the turning shovel, so that at the end of the turn the corn falls at the feet of the turner instead of several feet in front of him, so that when the whole piece has been turned, while the front edge may be six or seven feet from where it was before the turn, the back edge may be in practically the same position, and the whole piece is lying in considerably more room and consequently so much thinner. Incidentally it may be said here that if a piece has to be thickened by the turn, exactly the opposite treatment will be necessary, so that the turner will begin by throwing the corn high in the air and letting it fall close to front of him, and as he proceeds gradually making his turn a longer one till at the end of the turn he has gained the necessary number of feet.

To return to the piece under consideration. On Tuesday, the ninth morning out, exactly the same questions must be asked, viz., has modification progressed normally? Is the moisture content sufficient to keep it at work for the time necessary? Has growth been too fast during the night with consequent rootlet matting and tendency of acrospires to grow out? or has it been too slow, with undue drying of the corns, check to modification, and consequent loss of temperature? If the latter has been the case either the piece must be brought considerably together and carefully watched for undue rises of temperature, to be met by furrowing, turning, and, if necessary, thinning; or, if the check is very severe, it may be necessary to give a little more water and thicken up a little, in which case even more careful watching will be necessary, and loading will preferably have to be postponed till the twelfth morning if the weather and the pieces following allow of this being done. Under ordinary circumstances and when matters are going smoothly it will be sufficient to examine the pieces night and morning about this time before the customary hours of turning, but when pieces have got out of hand in any way it is most necessary that they should be carefully watched, as, especially after a check and subsequent thickening and sprinkling, temperatures may rise very suddenly, accompanying strong enzymicaction, which must be regulated as speedily as possible by aeration and movement.

To return once again to the piece under consideration. If matters go smoothly the number of hard-ended corns will have very materially decreased by the Wednesday morning (the tenth morning out,) while hard or pasty corns will be entirely absent, the temperature should not be above 65° F., and while individual corns will feel distinctly drier than they did, their contents, with the exception of the few hard ends, will rub down soft and mealy. By the Thursday morning (the eleventh day out) the hard ends should be entirely absent, and the piece should load perfectly mealy, but still fresh and without any marked drooping of rootlet or withering. Loading the pieces to kiln after a flooring period of ten days generally means, in fact, that the barley has malted easily, and that the whole process has gone smoothly with no undue waste of material, time, or labour. It is the shortest period in which barley can be malted as a rule, and though there are barleys which are capable of undergoing both complete and satisfactory modification in as short a period as nine days, these are so rare as to be the exception rather than the rule, and may be classed as those superfine qualities which must be beyond the reach of the every-day maltster and brewer, not the least of whose requirements must be to obtain really satisfactory material without paying the somewhat fancy prices which the very finest barleys must always command.

Before turning to the case of those barleys which require a longer time on the floors, a few words are perhaps necessary to explain the different temperatures at which the pieces are kept during the days between the finish of sprinkling and loading to kiln. Barleys vary so enormously in the way they behave on the floors, that it is impossible to write of the different treatment required by each separate variety, complicated as each case must be, apart from breed, by circumstances such as soil, manure, weather during the formation of the grain, harvest conditions, and subsequent conditions of sweat in rick, storage in barn, kiln-drying, and storage in the malting. Some barleys, as has been seen, will go through in ten days— of these some will require much more sprinkle than others— with all of them however, which require anything like large quantities, it is imperative that the desired amount should be added in a much shorter period than would be the case were they to be kept out on the floors for a longer period. The addition therefore of, in some cases, large quantities of water tends to promote a sudden increase of enzymic action, which in turn sends up temperature, and must be met by thin floors and constant turning. In the case of barleys requiring the longer flooring periods, which for that reason can have their water in smaller quantities, the increase of enzymic action and temperature is likely to be less sudden. These can, therefore, be kept thicker and need less turning, and even under these conditions will work at lower temperatures than the others. It is very necessary that they should do so in some cases, as working at higher temperatures during the prolonged flooring might be productive of mould during the last day or two on the floors.

Mould— The chief causes of mould in maltings are:

1. Dirt and dust.
2. Skinned barley and barley steeped with a quantity of damaged and broken corns present, or barley which bas been thrashed too close.

Both are aided by high flooring temperatures and a long flooring period.

To avoid the first, scrupulous cleanliness must be observed in the maltings, and the floors kept clean by being washed down with cold water and hard brushes or squeegees when possible. It is also most necessary to keep malt and barley dust off the floors both by arranging for any doors leading from them to the malt or barley stores to close automatically and to avoid the entrance of dust through the windows as much as possible.

Barleys which have been thrashed too close and those with the skin damaged from any cause should be avoided and great care taken when half-corns are present to see that these are taken out by the half-corn cylinders.

If these precautions are taken and the flooring method adopted is correct little mould should be present on the floors as a rule. The two varieties most commonly met with are:

Pencillium glaucum, a red or blue fungus first appearing on broken or damaged corns.

Asperguillus niger, which appears as a sort of white mist covering the pieces in bad cases, and is generally due to extreme dirt in the barley or extreme forcing.

The latter sometimes appears during the early stages of kilning where deficient draught and low temperatures are the rule.

The Long Flooring Period— Now to turn to a piece requiring thirteen days between emptying from cistern and loading to kiln—it has been emptied on a Monday morning and watered on the following Saturday at five days old, further doses of water being added on Sunday, Monday, Tuesday, and possibly Wednesday mornings. During all this time the temperature will have been kept between, say,

57° and 60° F. Two turns a day will probably have been given on Saturday and Sunday, but very likely furrowing may have taken the place of turning on subsequent afternoons. Now all this time modification will have been proceeding very much on the same lines as the ten days' piece, but considerably slower. On Thursday morning perhaps the activity may be at its height, when the temperature may rise to 62° F. This will probably entail a turn in the afternoon. On Friday there will probably be signs of fading in the rootlets, and the turn given will possibly be a very gentle one, followed by a rest in the afternoon by which time the temperature may have fallen to 59° or 60°. If there is any further falling off in temperature the piece may be slightly thickened up. Whether or not Friday afternoon and Saturday are set aside for the so-called withering process must depend upon the maltster. It is somewhat difficult to explain withering, but it consists essentially in stagnation of the rootlets accompanied by partial enzymic action in the corn, favoured by the dryness of the piece, and by letting the grain rest unmoved so that the carbonic acid gas formed can only get away very slowly. There should be no visible sign of sweat, although of course the liberation of C02 must be accompanied by liberation of water, apart from the small amount expelled by evaporation. Comparative dryness of the piece and low temperature are absolute essentials, and if either of these conditions is not observed a mild forcing process will take place, accompanied by growing out of acrospires, abnormal evolution of carbonic acid gas, and sweat. As an instance of this, a most obnoxious custom, less prevalent now than formerly, but still the rule in some places, may be cited, viz., that of thickening up the pieces before they are really rid of their moisture and letting them lie at high temperatures, almost heaped up for 24 hours before loading. This is, of course, far removed from the proper withering, the extreme of which is effected by the patent withering floor which consists of a perforated floor on to which the piece is turned some 24 to 36 hours before loading to kiln, and which is arranged so that a fan blows cold dry air through it, evaporation thus proceeding rapidly at low temperatures, say, 50°-60° F., while there is no check to enzymic action caused by temperatures approaching those at which any of the enzymes begin to suffer restriction of energy.

It has been found necessary above to mention definite temperatures for different stages of the flooring process, although this was not intended, as it is impossible to lay down any rules in this respect; if this were possible a temperature of 54° F. on emptying would be advised, rising gradually to 62° F. on about the fourth day, and then remaining fairly stationary at 60° F. until loading. Some barleys, however, require to be worked at distinctly higher temperatures than others, though as said before, 70° F. should never be reached. If pieces are allowed to work at consistently high temperatures during the first four or five days, there will be every chance that mould will be prevalent in them by the time they have been on the floors for eight or nine days, and besides this the loss due to unnecessary rootlet growth will probably be excessive. It is improbable, however, that excessive temperature and neglect of turning in the young pieces; i.e., during the first five days, ever produce what is known as "forced" malt, for although where forcing is the rule, the young pieces may often be neglected in those particulars, the real forcing treatment generally begins after the pieces have had their sprinkle, when high temperatures can be easily maintained by a combination of plentiful sprinkle, thick pieces, and infrequent turning.

Forcing— The object of forcing, of course, is to get the flooring process over as quickly as possible, so that more malt can be made in a house than would be possible on the legitimate system of flooring, but as really forced malt can be detected by its sickly sweet taste and excess of soluble ready formed carbohydrates, and soluble albuminoids, and the great majority of brewers will not have it, comparatively little forcing proper is now done, although, when barleys which are difficult to modify have to be dealt with, a sort of semi-forcing process may be sometimes resorted to in the hope that the malt may benefit in tenderness.

When malting operations have to be carried on in early autumn or late spring, the flooring process becomes more difficult, owing to the higher temperature and increased drying power of the atmosphere. The thinner varieties of foreign barley, by reason of their shape and size, lie lighter on the floors than English and the plumper foreign Chevalliers, and admit more air by reason of the greater interstices between the corns, and it is therefore customary to malt as much of them as possible during the warmer months of the season. As the higher temperatures at which the pieces have to be kept are likely to encourage the development of mould, it is as well to be especially careful that any barleys which have to be made up during these warm months should be particularly sound and free from chipped and broken corns. Apart from this, it must not be forgotten that the finer the quality of the barley the easier will its modification on the floors be, and, though it would be absurd to keep the finest barleys for warm weather malting, yet it is as well not to arrange to malt any barleys in warm weather which for one reason or another require either a prolonged flooring period or very generous allowances of sprinkle. Warm weather malting is never a very pleasant business; the pieces have to be kept thin to prevent unduly high temperatures, and this, of course, entails a large amount of surface exposure, and more evaporation, which is still further increased by the comparatively high drying power of the air. In maltings, furnished with sliding shutters as windows, and the form of curtains described above, this drying can be partially avoided by keeping the curtains damp by using a syringe every morning. Of course, in order to keep the pieces thinner more floor space will be necessary; and this will have to be arranged either by steeping smaller quantities of barley, or by increasing the intervals between the wettings.

It is especially important in warm weather malting that the rootlets should not be allowed to wither until at most twenty-four hours before loading, and it will generally be found best to sprinkle at four or three and a half days old in order to keep the young pieces fresh.

It is also as well to arrange as far as possible that the malt made in warm weather should be reserved for cold weather brewing, and vice versa. In normal seasons, November, December, January, February, March, and the first half of April may be regarded as legitimate months for malting, while during the latter half of September, October, the latter half of April and May flooring operations, at any rate in Midlands and south of England, become more complicated.

CHAPTER VII

KILN CONSTRUCTION AND DRYING

Drying and Curing— The subject of kiln construction is not an easy one. The malt kiln has to be used for two distinct purposes—drying and curing; for the former good draught is indispensable, and for the latter it is of relatively small importance. In a former chapter some attempt was made to trace the growth of the kiln from the small, low, heating chambers of a past generation to the more elaborate constructions now in use, and if reference is made to Figs. 1 and 6 it will be seen how much difference there is between them.

Kiln Temperatures— Now, although curing is extremely important, its success does not depend to the same extent upon construction as does drying, and it will be for that reason better to consider construction mainly with a view to the drying, as distinct from the curing, process. Drying malt consists essentially in passing large quantities of warm air through it, and the problem which confronts the designer is how most economically to heat a sufficient quantity of air to the required temperature. It has been found by experience that the temperature, as recorded on a thermometer stuck through the layer of green malt and resting on the kiln floor, but with its bulb protected from actual contact with the tile or metal surface, should not, under the most favourable conditions, register more than 130° F.; while under less favourable conditions of construction it may be necessary that the temperature, as recorded above, should not exceed 100° F. until a certain amount of drying has been done. The temperature taken as described above, is not, however, a very reliable indication of the temperature at which the air passes through the perforations in the kiln floor, which is, on a very rough average, some

30° F. higher.

Effect of Evaporation on Temperatures Recorded in Green Malt— Assuming then, that a temperature of 120° F. in the malt is aimed at, it will be necessary to supply air heated to about 150° F. to the under side of the kiln floor. In practice this is done by heating a small quantity of air by passing it through and over a fire, and lowering it to the necessary temperature by diluting it with a very much larger quantity either admitted to the shaft, where there is one, through draught holes, or, where there is no shaft, warmed by the air that has been in proximity to the fire, and carried up with it. Although the day of the shaftless kiln is over, there is a considerable distinction in the size of the shafts employed, and the old shaftless kiln may be taken as the extreme of the broad shafted kilns (Figs. 2 and 4).

Draught— Now it is obvious that, although all the air passing through the kiln is somewhat loosely termed draught, there is in reality a very real distinction between the air which has actually passed through and over the fire or furnace and that with which it is subsequently diluted.

Velocity— If a malt kiln be compared with the chimney of a boiler, and the two extremes of construction (Figs. 1 and 6) be taken, in the former case the whole area of the kiln is comparable to the chimney, and the whole body of air will be warmed to a small degree and ascend with a low velocity. In the latter case, however, only the air in the shaft will be appreciably warmed, so that a very much smaller area of air will be warmed and will ascend with a relatively higher velocity, until it reaches the top of the shaft and spreads out in the hot air chamber.

Assuming a kiln of 1000 square feet superficial area, with air passing up it to the extent of 5000 cubic feet per minute, the average velocity of that air as it strikes the lower side of the drying floor will be about 5 feet per minute. If, however, these 5000 cubic feet are taken up a shaft of only 20 square feet superficial area, the velocity of the air passing up that shaft will be increased to about 250 feet per minute, and the velocity through the small draught holes in the sides of the shaft may be as high as 1000 feet per minute.

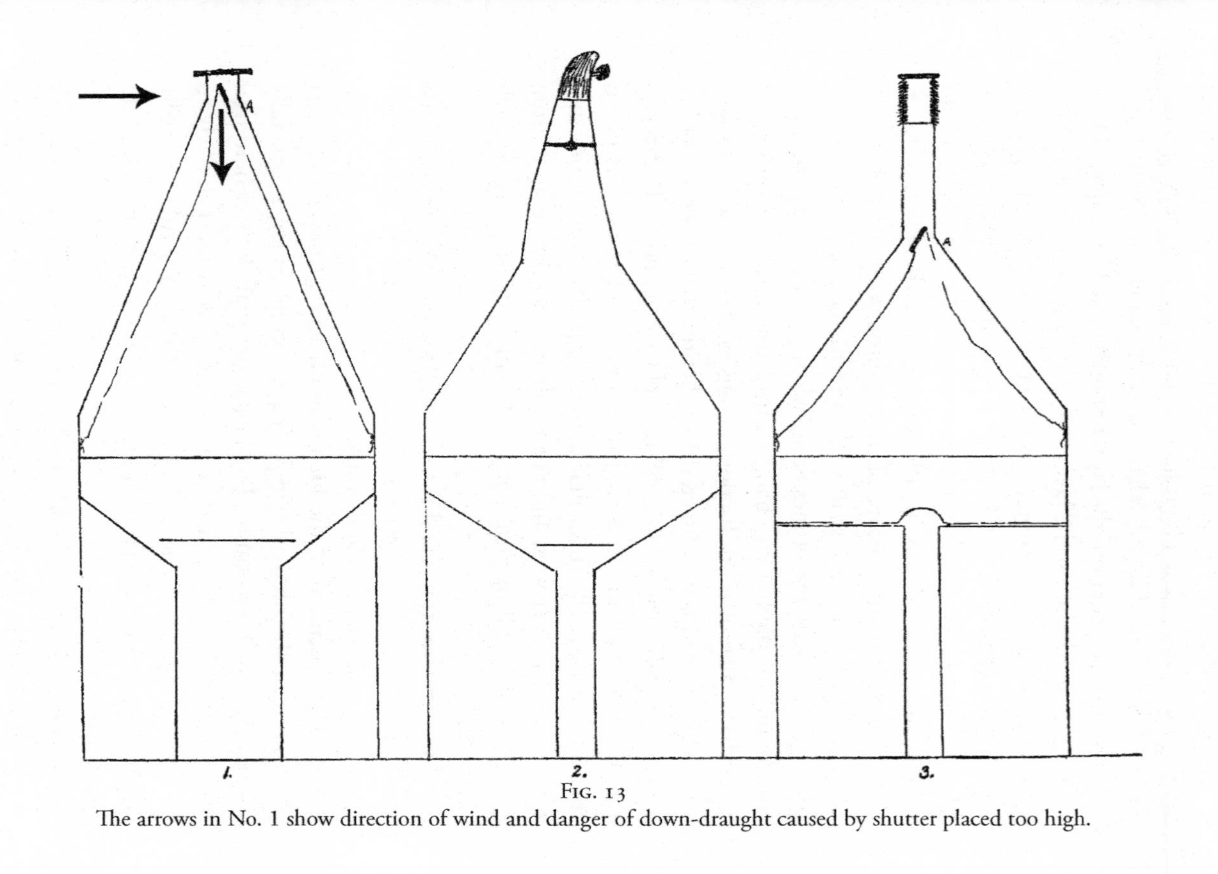

FIG. 13

The arrows in No. 1 show direction of wind and danger of down-draught caused by shutter placed too high.

Draught increases with the height of the kiln and the excess in temperature of the air in the shaft over the temperature of the atmosphere, but is diminished by resistance. In boiler chimneys the chief resistance is afforded by the relatively small opening of the fire grate which causes constriction, the layer of burning fuel, the curves in the flues, and the friction of the heated air and furnace gases against the sides of the flues and chimney; but in malt kilns by far the greatest resistance is caused by the layer of drying grain. There is also, in furnace kilns especially, the resistance offered by the fuel, but this only applies to the air passing through the fire. There is also the resistance caused by the passage of the air through the restricted areas of the draught holes, the shaft itself, and the octopus-like disperser in which the shaft sometimes ends, when it is necessary for equal distribution in the hot air chamber. In shaftless kilns or those with large shafts there is more friction, owing to the increased wall area with which the air is in contact; the smaller the shaft however, within reasonable limits, the greater will the velocity be, not only through the draught holes, but also, and this is the most important point, through the fire itself. This will lead to more perfect combustion and consequently keener fires, and as a result a higher temperature from the fire, which will in turn increase the draught sucked in through the draught holes.

In very many kilns combustion is incomplete. This, as is pointed out in E. S. BEAVEN's paper on "Fuel Consumption in Malt Kilns" (*Journal Inst. Brewing*, 1904, vol. x. No. 5), may occur either from excess of air passing over the fire or insufficient air passing through it. In kilns of the fire-basket type (Fig. 1) combustion is generally imperfect from both causes, but in narrow shaft kilns where iron doors are fitted to the furnace front the former is most unlikely to be the case, though the latter may be and often is the rule, especially when construction is at fault and the wind is not blowing right into the furnace.

Cowls— Down draught from the outlet of the kiln is a frequent source of trouble in some kilns, and probably the only real preventive of it is to fit cowls to the tops. These should be made preferably of copper and work on marble bearings, and where they are in use it is most important that there should be means of easily getting to them, as if they should get stuck during a change of wind they may become a source instead of a prevention of the trouble. Cowls working on ball-bearings have been introduced into some kilns lately, and with these of course the danger of the cowl not veering with the wind is greatly reduced.

Another point about kiln tops is that the outlets should be neither too large nor too small, the former tending to undue cooling, and the latter to unnecessary resistance, an evil which is also caused in kilns with mushroom tops by allowing insufficient distance between the cover and the aperture, and which may exist although the area of the outlet is sufficient.

Top Shutters and Down Draught— In the paper referred to above BEAVEN drew attention to the fact that it was much easier to regulate draught from the outlet than from the inlets of most kilns. It is somewhat difficult to prove that this is the case, as anemometer readings at the top and bottom of a kiln during the drying stages are not comparable owing to the amount of moisture which accompanies the air on its exit, and the difficulty of estimating the amount of air passing actually through the fire, but there can be little doubt that it is the case, and it is a great pity that it is so difficult to arrange a sliding valve at the outlet of a kiln. The constant change, however, from a dry heat to a damp heat makes it a matter of great difficulty, and it is hard to see how any practical sliding valve can be arranged in a way simple enough to make it proof against the constantly changing conditions. During curing, restriction of draught by means of the outlet is essential in some cases and beneficial in almost all cases, and it is very common to see a shutter fixed at the kiln top especially with mushroom top or louver top kilns. The simplest form of shutter works on an eccentric axis which leaves it open. It can be shut by a rope of copper wire, and it is of importance to arrange that the shutter when open should not have any part above the top surface of the outlet, as otherwise down draught is sure to be caused when the wind is at all strong, except of course in those fitted with cowls, where the shutter will be protected from the wind by the cowl itself.

Relative Position of Kiln Floor— As has been seen the velocity of the air passing through the fire is dependent to a very considerable extent upon the height and superficial area of the shaft. It is also dependent, and the total draught is more dependent, upon the total height of the kiln from the floor of the stokehole to the outlet. What is not quite certain is the effect of the relative position of the drying-floor. Taking the kiln of 1000 square feet superficial area, it would be necessary for the space between the top of the shaft and the drying floor to be about five or six feet in height to ensure good distribution and to allow of its being cleaned out periodically. To give head room at the sides of the kiln it would be necessary to carry the outside walls up at least four feet, so that if they were 28 feet in height the kiln roof would spring from the walls at a height of not less than 37 or 38 feet. Kilns vary greatly in construction, and it is very difficult to compare the relative efficiency of the different types, but in very many cases the distance between the drying-floor and the outlet is equal to or greater than that between the stokehole and the drying-floor, and though undue height leads to unnecessary cooling, it is probably better to err on that side, than to risk the greater resistance caused by a relatively flat roof.

Some outline sketches of roof construction are given on p. 87. In the No. 1 type there is a tendency to undue cooling owing to the large area above the drying-floor, and this is corrected in No. 2 by bringing the roof up at a less angle to a certain point and then carrying up a sort of chimney to the required height. In No. 3 this construction is carried still further so that the last 15 to 20 feet of the outlet take the form of a regular chimney of not much greater area than the fire shaft of the kiln, which is shown narrow in this case for purposes of comparison. The latter type does very well in some cases and it has the great advantages of simplicity and comparative cheapness of construction, while for curing, the flap at A can be closed quite tight; but there is, of course, rather more resistance than in the cases of No. 1 and No. 2.

Different types of construction beneath the drying-floor are also shown, No. 1 having the broad shaft and Nos. 2 and 3 narrow shafts. It will be seen that No. 3 shaft is carried higher than No. 2 shaft and is fitted with the octopus-like disperser which is necessary for good distribution when the hot air chamber is comparatively low and rectangular, as opposed to that of No. 2 whose floor is set at an angle from the top of the shaft; in this latter case the ordinary plate disperser is generally sufficient to ensure it.

It is not intended to suggest that the construction beneath the drying-floor and that above the drying-floor, as shown on the three kilns sketched, are necessarily found together in practice. Maltsters and architects have their own ideas on both points independently of each other, and as a matter of fact, where the first considerations are efficiency and economy in drying and curing, it is probable that the No. 2 construction beneath the drying-floor and the No. 3 construction above it would be the best, especially if the outlet were fitted with a cowl, which is in this case suggested on No. 2. No. 1 is shown with a mushroom top, and No. 3 with louvres; neither of these two latter tops is preventive of down draught, though the louvre type is to be preferred. A shutter is shown at A in Figs. 1 and 3. This would have to be at the exit in Figs. 1 and 2, and, when open, care should be taken that it does not protrude as shown in Fig. 1 above the outlet, or down-draught will certainly result in windy weather.

In a few recently constructed kilns uralite has been used for the ceilings, and it is probably the best material for the purpose as both wood and plaster are apt to cause trouble, unless the wood is very carefully oiled each season or the plaster very well put on.

In some kilns fans are used to increase the draught, but it is doubtful whether the benefits derived from them are sufficient to compensate for their initial expense, depreciation, and the power necessary to run them. Fixed in the top of the kiln they are very exposed to the effect of the steam, and, although when the kilns are constructed for them and their size is sufficient they, of course, do help draught, they cannot be said to have any marked effect on combustion, as the bulk of the air they draw must pass through the draught holes and not through the fire.

Contamination of Malt by Combustion Products— Since the arsenic episode of some eight years ago several attempts have been made to dry without passing the products of combustion through the malt, but they do not seem to have been very successful.

It is practically certain that the greatest absorption by the malt of arsenious oxide occurs during the earlier stages of drying while the grain is in a somewhat spongy condition; if arsenic is present in the fuel, and the temperature on the top of the fire at starting is low, there is every chance that the volatile arsenious oxide will ascend into the malt. The higher the combustion temperature the more readily will it combine with any bases present, and it is on this principle that Beaven's filter is based. To quote from his paper on the subject: "The furnace gases are carried at high temperature through a filter into which lumps of some basic substance, such as chalk, are placed. The chalk is very soon more or less converted into lime. Arsenious oxide, if present, readily goes into combination with the lime at high temperatures. Obviously the efficiency of the filter depends, like that of all filters, upon the proportion which the surfaces available for action bear to the volume of air passing."

This filter is in use in many kilns and has fully proved its efficiency, a great point in its favour being that it undoubtedly in many cases effects a reduction in fuel consumption, partly for the reason that it stores a large quantity of heat, so that where no night man is kept much higher temperatures are maintained during the night than would be possible without it. Of course the care necessary for the selection of the fuel need not be so great where the filter is used, and if it is not used there is no lack of coal which will give good results if care is taken to ensure as complete combustion as possible, and to keep the kilns, shafts, and hot air chambers well swept down.

Before going into the practical side of drying operations it may be as well briefly to consider exactly what has to be done during the drying, as apart from the curing, process. Although much emphasis has been rightly laid upon the different conditions necessary for drying and curing malt, yet there is a stage where the one merges into the other. Curing, that is the application of high temperatures without much draught, cannot be safely begun until the grain has been brought to a sufficient degree of dryness to ensure no harm being done to it, but taking a typical drying temperature at 120° F., and a typical curing temperature at 200° F., the difference is so great that it would be absurd to suppose that the grain, having been kept at the lower temperature until the expulsion of all but the last 5 percent of moisture, was suddenly raised to the higher temperature at this point. Curing proper, as a matter of fact, may be said to begin when the grain contains 4-5 percent of moisture, but, there is an intermediate stage when the malt although fairly dry to the touch, might be harmed by the application of high temperature such as 180-200°, but is unharmed by heats of, say, 140-160°, and when, as a matter of fact, the expulsion of the moisture at low temperatures would be a very lengthy process. The temperature permissible at any stage depends upon a variety of conditions, but it may be given as a broad general rule, that the heat permissible on starting should not be greatly exceeded until the moisture percentage is reduced to some 15 percent, when it may gradually be raised to the heat necessary for curing.

Drying proper then, may be said to end when the moisture percentage is reduced to about 15, and it will be as well to divide kilning into three stages, viz., drying, the intermediate stage, and curing.

Drying— When the old piece is ready for loading on to the kiln it may be regarded as a mixture of malt, water, and rootlets. Let it be assumed that for each 448 lbs. of barley steeped, 336 lbs. of commercially dry malt i.e., malt with a moisture percentage of 1 to 2 will be obtained, together with 12 lbs. of rootlets, or malt culms, of similar moisture content. The moisture percentage on loading may vary, as has been already stated, between 39 and 45 percent. Taking 42 percent as an average, and taking the moisture of the kilned malt as 2 percent on the average, 40 percent of moisture will have to be expelled. Then each 348 lbs. of commercially dry malt and rootlets will on loading constitute 60 percent of the mixture of green malt, rootlets, and water. If 60 percent constitutes 348 lbs., 40 percent will constitute 232 lbs., so that on the average, for each quarter of malt produced some 230 lbs. of water will have to be expelled during kilning. Similarly if the expulsion of only the first 27 percent of moisture necessary during drying proper is considered, some 160 lbs. of water will have to be expelled at comparatively low temperatures to reduce the moisture percentage to 15.

Thermometers and Recorded Temperatures— Temperatures

during kilning are generally taken on a thermometer pushed through the green malt on to the floor of the kiln so that the recorded temperature is that of the air passing through the grain, together with any heat evolved by chemical action in the grain (in some cases a substantial factor) and cooled by evaporation. The amount of heat unrecorded from this cause is very considerable indeed, and in order to read off the temperature of the air uncooled, several patent thermometers are in the market; the best of these is the thermometer enclosed in a double copper cover, as the air jacket between the cover guards the air round the thermometer from the cooling influence of the evaporation proceeding in the adjacent malt. Unfortunately, however, the very fact of the small space cleared on the kiln floor causes a rush of air to this part of the kiln, so that the velocity of the air passing round the thermometer is likely to be much greater than that at any other point on the floor; the bulb is also protected from the influence of any heat evolved by chemical action in the malt, so that while it must be regarded as a useful invention, it should not supersede the ordinary kiln thermometer. The simplest of these is probably the best as long as it is strongly made and the bulb and stem well protected, but when once a pattern has been selected it should not be altered without testing the temperatures recorded on the different patterns, as these sometimes vary considerably. A competent maltster who has been conducting kilning operations on the same kilns for some time, according to the suggestions of his employer, gets to know what he can do with his heats, under what conditions he may keep temperatures high, and when it is necessary to drop them somewhat; and although conditions must be so various on different kilns, and even on the same kiln under different conditions, that the actual number of degrees registered on a thermometer stuck in the drying grain and resting on the kiln floor can convey little or no definite knowledge, yet that figure combined with experience of former dryings, and, above all the amount of draught available and the condition of the grain, furnishes the knowledge requisite for successful drying.

Maximum Drying Temperatures Dependent on Many Conditions— The maximum temperature permissible during drying must vary chiefly with the amount of draught available, which in turn is determined by the height of the kiln, the heat from the fire, the condition of the atmosphere, and the amount of resistance afforded by the construction of the kiln and the load of drying malt. In the case of low kilns with deficient draught, much lower temperatures are necessary than on high kilns with good draught. In all kilns, during purely drying operations it is absolutely necessary, if economy in fuel consumption is to be ensured, that temperature and draught should be considered together. It has been pointed out that the temperature must be regulated to a certain extent according to the amount of draught available, which in turn is partly dependent on the amount of heat available from the furnace—if, therefore, a good natural draught is available due to kilns of sufficient height, skilfully constructed, a larger and fiercer fire will be permissible, and this in itself will tend to further increase the draught.

Saturation— It has been shown that during drying proper about 160 lbs. of moisture have to be expelled for each quarter of malt to be made. In the case of a 50-quarter kiln then, some 8000 lbs. or 56,000,000 grains of water are to be removed at comparatively low temperatures. Now the air to be used for drying will contain varying amounts of moisture per cubic foot dependent on temperature and weather, but its condition on entering the stokehole will be beyond the control of the maltster—its condition on leaving the grain loaded on the kiln floor, however, is very much under his control. To put the case as concisely as possible, the maximum amount of air which it is possible to saturate, or very nearly saturate, at the highest permissible temperature should be passed through the kiln. Supposing, for instance, the kiln of 1050 square feet of floor space with its load of 50 quarters lying at a depth of about 10 inches. Now, if the conditions are such that it is possible to pass 6000 cubic feet of air per minute through the load, a temperature of 130° F. may be permissible as recorded on the thermometer in the grain. This will correspond to a temperature of about 80° F. in the air above the grain; if this air be saturated, or nearly saturated, it will contain at this temperature some 9 grains of moisture per cubic foot, so that if the air in entering contains approximately 2 grains of water per cubic foot, each cubic foot of air will extract from the load some 7 grains of moisture, and as a result about 40,000 grains will be expelled each minute.

Dew-point— If, however, the temperature recorded in the malt were allowed to fall to 100° F., the air temperature would fall to somewhere about 70, and even if the same quantity were saturated at this temperature the weight of water absorbed by each cubic foot would fall from 10 to somewhere about 7, so that instead of 40,000 only some 30,000 grains would be expelled per minute.

Needless Excess of Air— If, on the other hand, it were possible to admit more air than could be saturated, although the amount of drying done would be as great as in the first instance, a certain amount of air would have been heated for which there was no necessity, and therefore waste of fuel would have been caused. Such indeed is the actual case in almost all kilns after the first hour or so, and in a great many kilns all through drying operations, and although it is very much better than the other extreme of cutting air-supply too low, it is a great pity that in most cases a great deal more air is heated than can be used for drying.

Hygrometers— The use of a recording hygrometer was advocated in the columns of the *Brewing Trade Review* over a year ago, and though it is not an instrument to which daily reference is advisable, it is essentially one which should be at hand in maltings to enable the maltster to see from time to time how nearly approaching saturation the air is during drying. The instrument is a simple one, consisting of dry and wet bulb thermometers and a movable pointer with a sliding index. By adjusting this to the wet bulb reading and bringing it to the intersection of curves from the two readings, the percentage of saturation and the number of grains contained per cubic foot of air can be read off on a chart fixed between the two scales.

Anemometer— An anemometer is another useful instrument when drying operations are going on; the velocity taken through the draught holes and multiplied by the area of the draught holes giving the amount of air going through the kiln, less the amount actually passing through the fire and over the fire where there are no furnace doors. The latter may be a considerable quantity, but it will generally be possible to arrive at only an approximate estimate of the amount of air passing actually through the fire. For instance, taking the case of a narrow shaft furnace kiln with doors closed over the fire and open beneath it, the orifice below the fire-bars being about 2 square feet in area, and supposing an average velocity in this space of 50 feet per minute,100 cubic feet of cold air would be passing through the fire each minute. If there were in the shaft four cold air inlets, each of 1 square foot and each recording an average velocity of 1000 feet per minute, 4000 cubic feet of cold air would be entering the shaft through them, making altogether a total of 4100 cubic feet of (cold) air passing through the hot-air chamber each minute.

Temperature, Velocity, Density and Resistance— Assuming the outside temperature to be 62° F., every additional degree of heat imported to the air would increase its bulk by 1/523, so that if the whole of the air were brought to an average temperature of 150° F. immediately below the green malt, for each cubic foot of cold air we should have 1 + 88/523 or, roughly, 1.17 cubic feet of air at the required temperature, or the original 4100 cubic feet, with a temperature of 62° F. and a density of 0·0761, would become nearly 4800 cubic feet, with a temperature of 150° F. and a density of 0.0650 (compare ROBERT WILSON on *Boiler and Factory Chimneys*, p. 3, published by Crosby Lockwood).

This question of velocity is an important one, as to a very great extent it determines the maximum temperature which can be safely maintained in the green malt. For instance, if the case which has just been cited were to be applied to a kiln with 960 square feet of drying-floor, the velocity through each square foot of floor would, if distribution were perfect, and the kiln evenly loaded, amount to 5 feet per minute, and with this velocity a recorded temperature of 130° F. would probably be permissible in the green grain very shortly after loading. If, however, owing to any circumstances the velocity through the grain only amounted to 4 feet per minute in each square foot, it would probably be unsafe to allow a higher temperature than 110° F. Why this should be so is not quite clear, but it is borne out in practice, and may be realised by a comparison of the initial heats used in high and low kilns respectively which constantly vary by 15° to 20° in favour of the high kiln.

Effects of Down-draught and Cooling at Exit— It must not be forgotten that the drying power of the air is determined by its minimum temperature while it is in the kiln, so that if from any cause the air after passing through the green malt is materially cooled before it emerges at the outlet its drying power will be materially decreased. The most frequent causes of cooling are undue size in the dome, and down-draught. If, for instance, the temperature of the air after passing through the green malt were 80° F. and it were 90 percent saturated, each cubic foot would contain some 10 grains of water, but if, before its exit from the outlet it were cooled down to 70° F., it would quickly become saturated at the lower temperature. Saturated air at 70° F. contains 8 grains of water per cubic foot, so that two of the original 10 grains would be condensed, and although a large proportion of the condensed water would probably go out as "reek "some would either fall back into the green malt or remain for a time on the ceiling in the form of dew, in either case, of course, requiring more dry air for its removal.

CHAPTER VIII

LOADING, DRYING, AND CURING

Loading— If kilning is to be a four-day process about two days will be taken up by the first stage of drying, previously referred to as drying proper, and during which some 25 percent of moisture is to be removed. This may be effected by either of two methods: (1) Loading all the old piece on one morning, (2) loading only half on the first morning, and having dried it sufficiently in 24 hours, turning it up round the sides of the kiln leaving nearly all floor space available for the drying of the second half, which is loaded on the following morning. This system is especially useful in low kilns and those with deficient draught, as during the time when draught is most important the resistance due to the layer of drying grain is practically halved owing to the reduced depth of the half load, and by thus diminishing the resistance and proportionately increasing the volume of drying air, higher temperatures can be maintained without danger of vitrification. Vitrification generally takes place in the bottom layer of the load which is exposed to the greatest heat both from the air and from the floor of the kiln, and results in a hardening of the corn, caused apparently by the action of the water. Whatever its cause the result is hard instead of tender malt, and in order to avoid it as much as possible where draught is deficient many maltsters take every precaution that the temperature should rise very gradually during the first four or five hours, until the bottom layer is dry enough to be beyond the danger.

In cases where the load of green malt is divided and loaded in two days there is good deal to be said for leaving the first half spread over the drying-floor, instead of heaping it round the sides of the kiln, and loading the second half on the top of it, though this mode of procedure is only possible on small kilns, or where swivel-spouts are used on larger kilns. If it is done the dry first half allows high heats to be used without danger of vitrification, as the damp malt is protected from the floor by the dry bottom layers, and the resistance caused by the latter is not very great, as it should be very fairly dry.

Furrowing on Kiln— It is customary after the kilns have been loaded for some hours to lighten up the grain, and where the load is a shallow one of 4 to 5 inches, this can be done by furrowing the grain with the stick-plough without moving the bottom layer from off the floor. Where, however, the load is a heavy one of from 10 to 12 inches forks have to be used, and with them it is impossible that some of the top layer should not change places with the bottom layer. This constitutes an additional reason for dividing the load where draught is at all deficient. For instance, supposing a kiln of deficient draught where the old piece is loaded on one morning to a depth of 11 inches, and danger of vitrification prevents the use of high temperatures during the first four hours, at the end of which period the kiln is forked over, then the temperature will still have to be kept low for another four hours or so until the grain forked to the bottom during the turn has become sufficiently dry to be beyond danger of vitrification. If, however, only half the old piece has been loaded, in addition to the increased draught and consequently increased permissible initial heat, the grain can be furrowed instead of being forked when advisable, and the temperature raised without fear of damaging the unmoved bottom layer now dried past much danger of damage.

Another advantage of loading in halves is that it splits up the work of the maltster. This is specially desirable where barley is steeped every five days and the flooring process is accomplished in ten days, as in these cases loading, sprinkling, and emptying the cisterns all fall on the same morning, constituting an exceptionally heavy morning's work once every five days, which is lightened to some extent when only halves of the old pieces have to be put on to the kilns. On the double loading system great care should be taken that only as much of the old piece as can be thoroughly well dried should be loaded the first morning, as if on the second morning it has to be turned up at all damp it will not be improved by lying in this condition for 24 hours while the second portion is being dried. On the third morning the portion loaded on the first day is turned down and mixed with that loaded on the second day, and both are further dried and cured together. It is probable that rather more fuel is necessary on the double loading than in the single loading system, and it is doubtful whether the former is ever very profitable in high kilns when draught is good. On the other hand, it is certainly most useful where kilns are low or draught is deficient.

Whatever the load and temperature, the air in the kiln should be kept as nearly saturated as possible during the whole process of drying. The temperature should never even in the best kilns exceed 130° F., and when anything approaching saturation is impossible at this temperature it is a pretty fair criterion that the moisture present has been reduced sufficiently to allow temperatures to be raised.

As has been previously stated, when the load is thin, stick-ploughs can be used and the malt furrowed, but when the loads are thick, forks have to be employed and the kilns turned. Whichever is done, the grain should be kept lying light, and generally at least three turns or ploughs a day are desirable during the first two days.

Kiln-Turners— While the load is damp it is not possible for men to turn the kilns with shovels, and the regular turning of the kiln load in this way is only done by kiln turners; these consist of blades fixed to a revolving shaft driven along the length of the kiln by means of a ratchet wheel engaging with a toothed girder. These blades are fitted with movable rubber ends which are capable of adjustment so that a clean removal of the grain from the floor is ensured. There can be little doubt that with kiln turners judiciously used a saving in both time and fuel can be effected, and as, if properly adjusted, they leave the load perfectly level, there is probably a slight advantage in regularity of curing effected where they are in use. With regard to economy, however, it is rather difficult to say whether they are desirable or not. Without doubt they require the kilns to be built with a view to their use, as long kilns are much better adapted to them than square ones. If this is done, and if when the general routine of the house is considered allowance is made for them, it is probable that they may be made to rather more than pay for outlay, depreciation, and power; but if they are installed in a malting unadapted to them they are very likely to be rather an extravagant piece of plant.

Double Floor Kilns— Kilns with double floors are out of favour in England. With them the upper floor is reserved for drying and the lower floor for curing, and there is generally some difficulty in arranging for suitable temperatures on both floors. For instance, while the bottom floor is being used for curing at a temperature of 200° F., and with little draught, it may be necessary to keep the top floor at a temperature of 100° F. with a good deal of draught. This generally means that it is necessary to admit air to the kiln between the two floors, where it is rather difficult to mix this air with the hot air which has done its work on the bottom layer of malt. It may be observed, however, that during curing in ordinary high kilns the velocity of the air above the malt is seldom much lower than 1 foot per minute, and taking an average curing temperature at 200° F. the temperature of the air above the malt is seldom less than 110° F., and it seems rather a pity that this large quantity of dry warm air should not be utilised.

Some fifteen years ago a good deal of malt was spoilt during the drying stage by vitrification and stewing. The latter process was due to the appreciation by maltsters of the necessity of high heats for the expulsion of water and their neglect of the accompanying necessity of a sufficiency of air. Thus, in starting drying it was often customary to shut off draught until the temperature in the kiln was considered sufficiently high. This led to vitrification in the bottom layer, and in the upper layers strong enzymic action exerted by the necessary conditions of heat and moisture; high percentages of soluble carbohydrate material and soluble albuminoids in the resulting malts ensued, and analysts were quick to realise where the harm was done and to advise maltsters accordingly, so that on starting drying operations it became customary to admit all available draught at the expense of temperature. This was all very well in the low kilns, but as larger houses and higher kilns began to take the place of the small original malt-houses it is probable that the desire for draught was carried rather beyond the need, with the result that larger volumes of air than it was possible to use were carried up through the loads, accompanied by low temperatures and therefore protracted periods of drying. The result was that instead of stewing there was a long period of strong enzymic action, due to the neglect of the maltster to find out how much air really was necessary and arrange his draughts accordingly. This is very often the case today, and it is a great pity that more attention is not paid to finding out the requirements of each kiln, for no irrefragable law can be laid down on the subject. Saturated air is a colourless gas, but when, owing to reduction of temperature, water is liberated a bluish grey reek is perceptible on the kiln. When this is perceived either the temperature should be increased or, if this is not allowable, more air should be admitted, and if this is not possible it is sometimes as well to clear a small space on the kiln floor to admit some warm dry air. Down-draught is the most frequent cause of reek, and it is seldom seen in kilns fitted with cowls.

The Stage between Drying and Curing— When the first 25 percent of moisture has been removed the temperature may generally be gradually increased until the malt is dry enough for curing operations to begin. This increase in temperature is effected, not by increasing the size of the fires, but by diminishing the amount of draught, and this should be done very gradually. On the four-day system the third day's work generally consists of reducing the moisture content from 15 to 5 percent, and it is probably at this stage that most fuel is wasted. It is very difficult to tell exactly how high the temperature may be safely taken without any danger of injuring the malt, especially as the moisture content of the load generally varies considerably in different places. It is most essential at this period that the kilns should be left perfectly level after each turn, as if they are not, the warm air making for the point of least resistance will most of it pass through the thinner part of the layer, drying this at the expense of the thicker portion of the load which will remain damp, and by its dampness prevent the application of temperatures sufficiently high to have much effect upon the drier portion of the load. This is essentially a part of the process where kiln turners are of very great assistance. It is probable that the drier the malt gets the higher becomes the temperature necessary for expelling the moisture remaining, and this fact may be easily realised on kilns with windows, where the glass remains dry at a certain heat, but becomes covered with "condense" subsequent to a rise in temperature, showing that this has expelled a further amount of moisture.

It is not easy to say how high the temperature may be allowed to go before the commencement of curing proper, but it is generally safe to let it rise gradually to between 165° and 175° F. If 180° F. is exceeded before the moisture percentage has been reduced to about 5 percent there will often be danger of colouring the malt, as it seems certain that malts take colour with greater rapidity when the moisture percentage is comparatively high. In fact it is very much better to err on the side of caution during the period immediately prior to curing, as even if the analysis of the resulting malt is satisfactory its palate may easily be effected for the worse by undue hurrying on of curing. During the last four or five hours before curing it is customary in many kilns, where flaps are used at the outlets, to partially close these in order to reduce velocity. This is a sound practice, for it must be remembered that as the malt dries it lies very much lighter on the kiln than when it is first loaded, so that resistance is decreased and a given area of draught inlet at the bottom of the shaft is likely to admit much more air in consequence. By reducing the opening at the outlet, however, much more effective control of velocities can be maintained, and it is for this reason that any practical method of reducing the outlet as required would be of such assistance during both drying and curing. As it is, by far the greater part of draught regulation has to be done at the bottom draught holes, and it is somewhat annoying to consider that the resistance afforded by the layer of drying grain diminishes as the necessity for draught diminishes.

Curing— When all but the last 5 or 6 percent of moisture has been eliminated from the malt, draught may be cut off and heats raised to the extent necessary for reducing the diastatic activity to the required degree and allowing the malt to acquire the colour and flavour desirable for the purpose for which it is intended. At he same time the moisture is reduced to between 1 and 21/2 percent, the lower figure being preferable if other considerations permit the somewhat strenuous conditions necessary for its attainment.

Tint— In the laboratory the tint of the malt is taken on the wort obtained from it, generally by the Lovibond tintometer, 1-inch cell and 52 scale; another scale with red and yellow glasses has been adopted in some laboratories, but is not so generally used as the older 52 scale to which maltsters and brewers have become accustomed. In his *Valuation of Malt* (The Review Press.) DR. MORITZ gives the average tint of a malt which has undergone no caramelisation as about 2½, and points out that in malts destined for pale ales the tint may vary between 3 and 4½, and in those destined for mild and black beers between 6 and 9. During the "drying off" or curing process the tint acquired depends upon a variety of circumstances, chief of which are the dryness of the malt when first subjected to high temperatures, the thickness of the husk of the barley, and the amount of air passed through the malt and the thickness of the load of malt on the kiln floor. The degree of modification attained in the green malt has also a certain influence on tint, tender malt taking colour more easily than hard malt.

Turning During Curing— In order that curing throughout the load should be as even as possible it is advisable that one, two, or three turns should be given during the process. This turning is not quite the same thing as turning on the floors. In the latter process the turning shovel is allowed to cut through to the floor and the grain is swung either forward or sideways into the air by a heave of the shoulders, a turn of the wrists as it leaves the blade of the shovel causing the portion which lay next the floor to form the top layer of the turned piece and vice versa: aeration and change of position are thus ensured for the growing grain. But in kiln turning the latter only is necessary, aeration and consequent cooling being avoided as much as possible. In kiln turning two strokes should be made, the proportionately greater depth of the kiln load making this possible. With the first stroke only the upper portion of the load is lifted and turned over on to the empty floor in front, while with the second stroke the malt which lies next the floor is lifted and deposited on the top of that lifted at the first stroke. This chopping over with the double stroke is also sometimes used for moving couches in cold weather, when too much aeration is inadvisable owing to the loss of heat entailed thereby.

Caramelisation— During drying off, caramelisation and consequently colour is apparent first in the layer of malt in contact with the floor. When the maltster from an examination of the bottom layer considers sufficient colour has been formed, the kilns are turned in order that the upper portion of the load may be cured to the same extent. On some kilns one turn only is necessary, on others two or even three. The greatest practical aid to colour estimation is afforded by the culms, and though this is only a very rough test, it may nevertheless be made to serve its purpose if intelligently carried out. For instance, it may be observed by experience that in a given kiln, a given shade of culms from a thick-skinned barley such as an English Goldthorpe or a Brewing Chilian corresponds with a given tint in the resultant malt; while a similar culm tint from a thin-skinned barley, such as an Archer or a Brewing Californian, means rather a higher tint in the wort. Where the malt is destined for a brewer who is particular about the tint of the malt, the best plan is to take handfuls of malt from the kiln floor when necessary and sieve them, so that the colour of the malt may be roughly estimated and the malt turned or not as necessary. The test, of course, is a very rough one, experience of the tints of previous kiln loads of malt made from similar barley being indispensable. It is also absolutely necessary if the culm tint is to have any value, that high heats should on no account be employed until the moisture content is reduced to about 5 percent. In practice the dryness of the load is estimated by the "tread" of the kiln, i.e., by walking over it, when experience of former dryings gives a very good idea of the comparative dryness of the load. Thus, if the malt treads "crisp" and the feet sink into it, it is generally a very fair indication that curing heats may be allowed, while if it treads "dead," and there is a soft patch here and there, it is a sign that another turn should be given before heats are raised. If this is done before the malt is sufficiently dry the colour will be excessive, malt when damp taking colour more readily than when dry.

Effect of Withering on Colour of Rootlet— Another point which should be observed is the state in which the green grain is loaded to the kiln. If it loads fresh, the culms will be perceptibly brighter than will be the case if it has been exposed to a long withering process, while if much mould has been allowed to form, the culm test will be practically useless, as the culms will be of a dingy colour. If the malt is to be dried pale, no red rootlets should be perceptible, the shade rising from pale amber to bright canary as lower or higher tints are desired. If coloured malt is wanted a proportion of red rootlets is permissible, but they should be bright red, as dark red culms generally mean that caramelisation has gone too far. It is probable that DR. MORITZ was the first to point out that excessive caramelisation means loss of extract, and that this begins when a tint of 9 or 10 is exceeded. When this is allowed to occur the starchy content of the caramelised corns are a light brownish yellow instead of white, and if a badly caramelised sample is placed in the cutter, a large percentage of the corns will be seen to be of this shade when cut transversely.

Flavour— Over caramelisation may be easily detected by the palate of the malt, and once this flavour is recognised, over caramelised malt can always be detected; apart from this, the flavour of the malt is most important, and should be a serious factor in the estimation of its value as brewing material. A full sweet flavour should distinguish malts cured for mild ales, while a clean dry delicate palate is required in those to be used for pale ales. When the flavour is sickly sweet, it is generally an indication that the growth has been more or less forced; but this somewhat mawkish taste is often covered by high curing heats and partial over caramelisation when the maltster is cunning. When there has been mould on the floors, a musty flavour may be traceable, but this also, unless very excessive, may often be covered by a little extra curing, and careful-screening. Every maltster who is in charge of curing operations must have his standard of flavour to help him in his decisions on the time when turning is necessary, and the amount of temperature and time desirable.

Diastase— In the paper by MR. GORDON SALAMON on malt making, previously referred to, some Tables were given tracing the growth and decline of diastatic activity during flooring and kilning; these figures were obtained from analyses of malts made from high-grade English: barleys of the same harvest, malted apparently on somewhat different systems, and a reference to them will show that the diastatic value of the malt calculated to 2 percent of moisture varied, at the time the grain was loaded to kiln, between 126 and 66 Lintner, and in some cases rose and in others fell after some hours of kilning. It is probable that if malt is loaded very fresh to kiln, say, with 45 percent of moisture, and exposed to a gentle heat, say, 80°-90 F., there will be a rise in diastatic power, while if loaded in the same condition and exposed to high heats of 120-130° a fall in diastatic power will be the result. After some 48 hours kilning in the cases in question the diastatic activity varied between 60 and 36.

When it is considered that these very divergent results were obtained on sound malting systems and with high-class barleys of the same country and harvest, it may be realised how impossible it becomes to give any information which is likely to be of the least value on the far wider question of different races of malting barley of various harvests and all parts of the world. With regard to the malting process it is pretty well ascertained that long steeping periods, plentiful sprinkle, long flooring periods, and comparatively high flooring temperatures tend to give high diastatic value at loading; thus, barleys which can be malted with short steeping periods, little sprinkle, low temperatures, and short periods of germination, are likely to load with low diastatic energy. Some barleys have the reputation of being easily maltable in most seasons, such being the highest-grade English, Hungarian, German, and Chilian Chevalliers, Brewing Californian, Ben Ghazi, Spanish, and the finer qualities of Cyprus. Others, such as the coarser English Chevalliers and Goldthorpes, Ouchacs, Brewing Chilians, and Smyrnas, often require a more strenuous malting process and are likely to load with high diastatic activity. But conditions of field ripening, harvest, and subsequent maturation in store, also have far-reaching effects on the question. Barleys from the 1905 harvest were notably strong in diastatic power, though many of them malted very easily, and it was a very difficult matter in the 1905-06 season to effect the desired reduction on kiln without over-colouring the malt for pale beers. The barleys of the 1906 harvest, however, came to hand somewhat unmatured in many cases at the commencement of the present season, and many maltsters found a great difficulty in making a quantity of diastase on the floors sufficient to resist the curing heats necessary, although a long flooring period and rather forcing lines were adopted; this want of diastase, however, was accompanied by deficient extract and low sugars, and ceased to occur after the barleys had been in store for a month or two.

Efficient and regular control of diastase can, in fact, only be attained by experience of barleys, flooring methods, and kilning methods, and no cut and dried rules for curing can be given. In some seasons a final temperature of 200° F. for eight hours may be necessary to determine a diastatic capacity of 40 Lintner, while in others only 180°-190° may be necessary. In practice, tints of from 3-41/2 combined with diastases of about 40 are generally required for pale, and tints of 6-9 with diastases of between 25 and 30 for mild ale malts. Where low tints combined with high diastases are required, proportionately more air must be passed through the malt during drying off and more frequent turning given. Where the green malt is loaded to kiln with a very high diastatic value, and it is desired to bring this down to 25-28 without undue colour, it is sometimes advisable to reduce draught as much as possible directly the malt is dry enough to make this safe, and run the heats up to the limit where no colour is formed, say, 160-170°, for several hours before actual curing begins.

Rounding up on Kiln— Some maltsters make a regular practice of curing to a point rather below that desired, and then rounding the malt into large heaps for some hours before it is thrown off the kiln, with the object of ensuring regular curing throughout the load, and in kilns with bad distribution of heat this is often a good thing, though where this is not the case it is not necessary and in some cases may lead to uneven curing, the heat being naturally greatest on the thickest part of the heap and proportionately lower at the outsides.

If it is desired that the moisture content of the finished malt be very low (1 percent or under), extra care should be taken that high hats are avoided until the moisture content is reduced to 4 or 5 percent, and even then the draught should never be completely cut off either at the inlets or at the outlet, for though the latter is of course practically impossible in most kilns, yet in those furnished with outlet flaps there is a good deal of difference in velocity effected between shutting the flap as tight as possible and leaving it open 3 or 4 inches.

Moisture and Redrying Slack Malt— In some maltings it is customary to withhold a little malt as each heap or bin is finished, and to redry the slack malt with subsequent kilnings of similar material; where this is done, and the malt to be redried has been screened, care should be taken that only a small proportion, say, up to 5 percent, be mixed off with each kiln load, as if large quantities are added there is a tendency for the screened malt to cause the whole load to lie rather heavy on the kiln, with the result that draught is reduced and colour is taken more readily by the bottom layer. If this is done it is better to let any malt to be redried form part of a kiln load than to wait until there is a sufficient quantity to be redried by itself; this for two reasons. First, slack malt does not improve on keeping, but deteriorates in value, and should therefore be redried as soon as possible; secondly, because if the malt has been properly cured in the first instance it will be a somewhat difficult matter to expel the moisture present without unduly crippling the diastatic energy. It is rather doubtful whether the last 4 or 5 percent of moisture can be eliminated at any heat under 180° F., and if malt with a low diastase is to be redried at this temperature there will often be danger that after redrying, its energy in this respect will be so crippled that it will only be fit for blending off in very small proportions in malts destined for running beers or porter.

For the same reason it is most advisable at the beginning of a season, or when a new lot of foreign barley is to be cured, to err on the side of under curing rather than over curing.

Fuel-consumption— The amount of fuel consumed during the drying and curing process varies enormously according to the construction of the kiln, and the method employed. Before the detection of arsenious oxide in malt, it was customary in many maltings, especially in the north of England, to use gas coke either alone or mixed with oven coke or anthracite. In the cases where gas coke alone was used, it was sometimes the rule to set sale of malt culms against purchases of coke, and in very many cases these items of revenue and expenditure balanced each other. Now, taking yield of culms at 14 lbs. to the quarter of barley steeped, and pricing them at £4 per ton, the return from this source works out at about *6d.* per quarter, or with gas coke at 10*s*. per ton, allows 1 cwt. per quarter for drying, curing, and barley sweating.

The latter is a very small item, and an estimate of 1 cwt. of fuel per quarter of malt made would probably be very near the mark for small kilns before gas coke had to be abandoned for drying purposes. The same figure holds today in certain districts, even in the case of high kilns, but in others it has been very greatly reduced, and there are probably kilns today where the fuel consumption is as low as 45 lbs. per quarter.

These however are exceptions, and the more usual figure is from 58-80 lbs. Basket fires consume more fuel than furnace fires, and in my opinion, no well-constructed furnace kiln should consume more than 60 lbs. and no well-constructed basket kiln more than 75-80 lbs. per quarter. It has sometimes been asserted that malt from foreign barleys takes more drying than English. This is a misapprehension caused by the relatively greater increase of the former. Thus, if 100 quarters of, say, Brewing Calfornian barley are steeped, and 100 quarters of English are also steeped, the foreign will take more fuel than the English; but for the very simple reason that, while about 107 quarters of Brewing Californian malt will be made, only some 101 quarters of English malt will result when both malts have been weighed up to the quarter of 336 lbs.

Arsenic— Most brewers now require that the malt should be accompanied by a certificate from some competent analyst that the amount of arsenious oxide present does not exceed a certain proportion. Standards vary in different breweries, but the standard of purity recommended by the Royal Commission on Arsenical Poison was 1/100 grain arsenious oxide per lb. for solids, and few breweries are content with a higher proportion than 1/350 to 1/400, while others demand that only 1/700 or 1/800 should be present.

Fuels— If no precaution beyond ordinary care and cleanliness is adopted and the combustion products of the fuel are passed directly through the malt, it is necessary that great care should be taken that only fuel comparatively arsenic-free should be burnt. The highest standard of purity is probably attained by South Wales anthracite, certain collieries being very much superior to others in this respect.

Well-burned oven coke is also in some cases of remarkable purity, but gas coke is almost invariably a dangerous fuel to use. The price per ton of anthracite and oven coke varies very greatly with the state of the market and the position of the maltings, the oven coke being in some cases the more expensive fuel of the two.

As regards actual use, however, the best oven coke is probably more economical than the best anthracite, both because the coke lies somewhat lighter on the fires and is capable in many kilns of more complete combustion. Also because anthracite is naturally very brittle, and in the course of transit from the colliery, unloading at the maltings, and even during stoking, a good deal of it gets broken up very small, these small pieces and the dust having to be reserved either for banking the fires at night or for being worked off in small quantities with the larger lumps. If too great a proportion is used dull fires are the result, except in the case of kilns with exceptional draught, and the constant poking necessary entails a considerable loss through the fire bars; this is not the case with coke, which burns to a hard clinker and none of it is wasted in the ash-pit.

Oven coke, however, gives rather a fierce fire, and in most cases it is advisable to use at any rate a proportion of the best anthracite for curing and for night work.

Stoking— There are many methods of stoking in vogue, due in the main to the many different types of fire baskets and furnaces in use. With respect to fire baskets, these are generally piled up with a mound of fuel, which takes some time before it becomes well ignited, combustion proceeding from the bottom towards the top of the mound; this causes a comparatively long period of low temperatures on the outside of the heap of fuel, helped by the considerable volumes of cold air ascending round the fires. These low temperatures, as has been pointed out in a former chapter, favour the escape of any volatile gases present, such as arsenious oxide, into the malt during the long period of incomplete combustion in the mound of fuel. When once this mound is well ignited it does not require much attention for a long time, but after fires have been made up, rather a long time must elapse before the full effect is felt in the kiln.

Fire Baskets and Furnaces— With furnaces, however, it is customary to keep comparatively thin layers of fuel only, and when these cover all the fire bar area and the furnace doors above the fire bar level are closed, it is obvious that little air can ascend round or over the fire, but that most must pass through the thin layer of fuel. This entails far more constant stoking, but on the other hand the kilns are warmed a good deal quicker and combustion is much more complete. On loading mornings it is an almost universal rule that the fires are not made up until the kilns are loaded. While a basket fire may take an hour and a half to become well ignited most well-constructed furnaces may be made to burn through in twenty minutes, so that the time saved in this way is quite appreciable.

CHAPTER IX

BARLEY AND MALT

Effect of Barley on Resulting Malt— It has been attempted to show, in preceding chapters that the difficulties encountered during the manufacture of malt on the flooring system are relative to the ease with which the barley selected can be made to adapt itself to the malting process as carried out on that system, and though this may appear on the face of it too obvious to need emphasis, a closer scrutiny into the relative values of barleys and malts will show that the matter is not, in reality, quite as simple as it seems.

Pale Ale Barleys— It will be best to take the case of malt first, and in estimating its value it will be necessary for this purpose to assume that the drying and curing processes have been well and successfully carried out because, though of great and even vital importance to the subsequent value of the malt, their successful issue does not depend in the same way upon the barley as does the flooring process. The malt, in fact, is virtually made when it is loaded on to the kiln, and though bad kilning may spoil it, no competent maltster is ever likely to make this mistake once he has realised what is required. The acquisition, then, of the desired tint, the regulation of diastatic activity to the required extent, and the expulsion of a sufficient amount of moisture without vitrification, caramelisation, scorching, or "magpieing" may be taken for granted. The flavour, however, although dependent upon the correct regulation of heat, draught, and moisture, must also be influenced to a very considerable extent by the peculiar characteristics of the barley employed, and, assuming that the barley has been freed from dirt, dust, seeds, very thin corns, and broken corns, a great deal must depend both directly and indirectly upon the quality of the raw material selected. Thus the delicate flavour required for pale ale malts is only attainable in its perfection from the finest qualities of sound clear-grown Chevalliers, in which category must be included the Hanna and Archer varieties, though, strictly speaking, these are variations of the true Chevallier type. The flavour of the malt in these cases is not only inherent, so to speak, in the barley, but is also the outcome of the ease with which these fine barleys adapt themselves to the malting process. Long flooring periods, liberal allowances of sprinkle, warm flooring temperatures, and constant work, either with the turning shovel, fork, stick-plough, or rake, all tend in greater or less degree towards broken skin, vigorous growth, and strong enzymic action, resulting in high percentages of soluble, ready-formed carbohydrate and albuminoid material and liability to infection by moulds. The fine barleys referred to being easily maltable without these conditions, and escaping the dangers engendered thereby, produce malts whose flavours are unimpaired by such causes.

Running Beer Barleys— When, however, the malt is destined for running beer, Burton beer, or strong old ale, delicacy of flavour is not so essential as the acquisition of a clean, sweet, full palate, dependent to a far greater extent on a carefully controlled curing process, and although mould in growing pieces is as undesirable as ever, the result of its presence in small quantities is far less felt in coloured than in pale malt. Again, with regard to the flooring process the conditions necessary for the successful modification of the somewhat coarser types of barley, such as relatively high temperatures, plenty of sprinkle and vigorous enzyme action, are not necessarily ill suited to the acquisition of the palate required for coloured malts, always provided of course that the flooring process adopted justifies itself by turning out malt satisfactory in respect to plumule and rootlet growth and percentages of soluble carbohydrates and albuminoids.

It therefore appears that while for pale ale malts, barleys which malt easily are essential, such is not the case for mild ale malts, as long as the barley can be satisfactorily made up without the evils attending undue forcing, and so long as it is of fair size and quality. It is not intended that the deduction should be made from these remarks that any barley which will grow and modify is good enough for the manufacture of coloured malts; certain barleys such as very thin English, most Danubians, and a large proportion, of the coarser types of six-rowed barleys such as Algerian, Tunisian, and the French and Belgian "Escourgeons," produce malts which very seldom acquire a good flavour, however skillful the flooring and kilning processes may be, and this may or may not be due to the high percentage of some form of albuminoid matter often present. With English barleys, quality and maturation being equal, the largest generally make the best coloured malts with regard to flavour, size seeming to give a certain "body" or fullness to the palate which is seldom obtained from the smaller sorts, and it is for this reason that the bold Goldthorpe barleys of Scotland, Ireland, Yorkshire, and parts of Lincolnshire, are such favourites for coloured malts when the seasons are suited to their development, ripening, and maturation.

Brewer's Extract— The question of extract is a most important one, as upon the amount of extract obtainable must depend in part the commercial value of the malt, provided that it is satisfactory in other respects. Unfortunately there has been a tendency lately to value malt almost exclusively by its extract, and to neglect other matters, which, while less easily discernible, are none the less important in their bearing on the type of beer ultimately produced. In fact the tendency is to overestimate the value of the quantity and to underestimate the value of the quality of the extract obtained. This is all the more to be deplored, because as MR. A.C. CHAPMAN has pointed out, the expression "the extract" of any malt is an incomplete one. Any given malt may be made to yield extracts varying according to the fineness of the grist prepared from it, and in order to create a basis for the comparison of the extracts of different malts it is necessary that each malt should be ground in the same type of mill, and at the same setting. This basis of comparison has been actually afforded by the Malt Analysis Committee of the Institute of Brewing, the standard grind being that afforded by the Seck mill set at 25°, this being considered the nearest approach to the average grind actually used in breweries.

Grinding— It now becomes necessary to consider two points bearing on extract— (1) degree of modification attained, and (2) size. (1) The farther the modification of any barley has been carried in the malting process the coarser may be the grind necessary for obtaining the maximum yield of extract in the mash. Thus if two malts, one completely and the other incompletely modified are ground at, say, the 40° setting of the Seck mill, the difference in favour of the former will be far greater than if the grinding had been carried out at, say, the 20° setting of the same mill. This fact has led some analysts to compare the extracts obtained by crushing the malt and grinding it at the close setting. If the settings adopted are regulated by the size of the malt this method gives the buyer all the information he requires. For instance:

Malt "A" ground at 25 Seck yields 94 lbs.,
Do. crushed at 40 Seck yields 93 lbs.,
Percentage crushed to ground 99;
Malt "B" ground at 25 Seck yields 94 lbs.,
Do. crushed at 40 Seck yields 90 lbs.,
Percentage crushed to ground 95·9,

the conclusion being that malt "A" has been well and malt "B" less well modified, and that while the extract actually obtained in the brewery may very likely be the same in both cases; *malt* "A" will be the superior article, and as such possibly better fitted for the production of certain qualities of beer than *malt* "B."

Crushing and Grinding— (2) It is obvious that if malt is to be crushed, and not ground, the setting of the mill must depend upon the size of the malt, size in this case being synonymous with relative plumpness. Thus, if a large "*cobby*" barley, such as a Yorkshire Goldthorpe, be ground at the same setting as, say, a thin Gaza, and that grind is such as to crush the larger malt, many corns of the thinner will escape untouched by the rolls, so that if extract is to be taken as a means of ascertaining relative modification it will be necessary to set the mill differently, according to the size of each malt. As a matter of fact; the standard 25 setting of the Seck mill is such that very few if any corns of even the thinner varieties of foreigns escape uncrushed, but the very fact that the standard grind is such a fine one makes it obvious that the extract alone, as taken by that grind, cannot be taken as any sort of standard of modification. It may be said that a brewer does not want to be able to deduce relative completeness of modification from laboratory extract, but to be informed as nearly as possible what extract he will obtain in practice in the brewery, and that this object is achieved very fairly well by the standard grind. This may be all very well in many cases where the brewer is a good judge of malt, or where he is quite clear on the real facts bearing on the question; but it is very necessary to be quite clear on the point that a comparatively high extract obtained from a finely ground malt does not necessarily mean that the malt has been well modified, and conversely that a comparatively low extract, caused either by a small malt or crushing, as opposed to grinding in the mill, is not necessarily an indication of either incomplete modification or want of judgment on the part of the barley buyer.

Setting aside the constant change in the setting of the rolls necessary for the estimation of comparative modification in malts of different size, and the estimation of percentage extract crushed to ground malts, neither of which unfortunately comes quite within the range of practical everyday work, two methods present themselves for the estimation of relative modification both of which are very simple and useful to the maltster and brewer, but which are both affected by the personal equation. They are (1) the "bite," and (2) the specific gravity of the malt.

The "bite" of Malt— The determination of the relative modification of a sample of malt by biting it down is largely a matter of experience; in general, however, it is important that one or two points should be remembered before expressing an opinion. Hardness may be due to several distinct causes. For instance, if the barley contains a proportion of dead corns it will be easy to trace this by biting, or if the whole piece was distinctly undermodified on the floors, it will be equally easy to judge. In both these cases the malt may be rejected as containing too large a proportion of insoluble starch. If, however, the malt is very slightly undermodified, each corn having the extremity of the distal end just hard, it is very difficult to realise this by biting, the little fragments of insoluble starch being so small as often to escape notice, although if a perfectly modified malt were bitten immediately before or after, the difference in bite might be quite appreciable, Again, the thickness of the husk must be taken into consideration, especially in foreign malts; thus it would be absurd to expect a typically thin-skinned brewing barley, such as a Ben Ghazi, to bite down like a typically thick skinned barley, such as a brewing Chilian. Although the starch content might be equally well modified in each case, the tough skin of the latter would undoubtedly influence the opinion of the biter unless due allowance were made for it; and this is also the case to a lesser extent in English, the thick-skinned barleys, such as Goldthorpes, never biting quite so tender as the thinner skinned varieties of Chevalliers.

The "Sinker" test— With regard to the specific gravity of malt, MR. E. S. BEAVEN, in his "Varieties of Barley" (*Journal of the Institute of Brewing*, vol. viii. No. 5, 1902), states that "the specific gravity of malt (immersed in toluol) varies from 0.95 to 1.1," and, "generally speaking, malts with specific gravity of less than 1.0 are well modified. Malts with specific gravity of over 1.0 may or may not be well modified." This statement has not been verified by the actual experiences of the writer, but from a long experience of the way malts float in water it is tolerably safe to say that well-modified English malts, when fresh off the kiln, but of course cooled to the normal temperature, should not show more than 5 to 6 percent of sinkers. With thin-skinned English barleys of the finest sorts sinkers are very often entirely absent, and even with fine qualities of the coarser skinned sorts all the corns very often float. With certain varieties of foreign, notably Brewing Chilians and Syrian Tripolis, the percentage of sinkers is generally a good deal higher, even when the malt bites down well and the percentage of extract of malt crushed to malt ground is high.

The sinker test, in fact, though primitive, is an extremely useful one for the guidance of the maltster who has not time to analyse every sample in the laboratory. Not less than 200 corns should be taken, and the sinkers should be examined to see whether their weight is due to dead barley, vitrification, or hard ends; the latter being generally easily picked out owing to their standing up vertically on the bottom of the glass, hard end downward. Of course both the biting test and the sinker test depend to a considerable extent upon the moisture percentage in the malt being low, say, not above 2.5 percent, and when malt is bitten down to ascertain roughly its degree of slackness more attention should be paid to slight apparent slackness in thick-skinned and thin foreign barleys than in English which is never quite so dry to the bite even when the actual moisture percentage is low.

Appearance of Malt— It is a great pity that so much importance is attached in some quarters to the appearance of malt. Apart from the presence of mouldy, damaged and magpied corns, or scorched corns, the actual brightness and relative attractiveness of the appearance of a sample very seldom affords a reliable criterion of its actual brewing value. The old proverb that appearances are deceptive is perhaps, as true of malt as of anything. More frequently than not hard malts are superior in appearance to tender malts, and it is most necessary that appearance should only be considered in conjunction with analysis and flavour.

Pale Ale Barleys— It might fairly be assumed from the stress laid upon the desirability of special qualities in the barley to be selected for the manufacture of pale ale malts, that its choice was a more difficult matter than that of the material to be used for mild ale malts. Such actually is not the case. On the contrary, the recognition of the finest lots of barley is a comparatively easy matter, once the buyer has realised the outward signs for which it is necessary to look; any one with a fairly well-trained eye can easily detect in barley thin skin, evenness in size and colour, and freedom from corns broken or skinned during thrashing. Almost as easy is it to recognise barleys which will malt easily, although it is almost impossible to describe the appearance which gives this promise. Delicacy and crinkle in the skin are generally present, but above all, the appearance and colour of certain English barleys tell the tale of ready maltability to the practised eye. When, in addition to this appearance, the barley looks bright and pleasing, smells clean, feels dry to the hand, and, when cut or bitten transversely, shows a white, mealy endosperm, the promise of easy maltability is seldom or never falsified.

The very fact, however, that the finer qualities of barley are so easily recognised assures their market value to a very great extent. There is always a demand for them and, though their market value may fluctuate slightly according to the laws of supply and demand, they are very seldom to be bought cheap.

Now in purchasing barleys destined for the manufacture of high-class pale ale malts, it is absolutely necessary to select those which show that they will malt easily, but there are certain qualities of barley from all countries which, while showing none of the usual indications of ready maltability, are nevertheless capable of being made to turn out very good malt when skilfully treated in the malthouse. In fact, while any barley buyer worthy of the name can tell what barleys are practically certain to malt easily, very few have the power of selecting those which, while lacking the external appearance generally associated with quality, will nevertheless be capable of being easily malted.

The most difficult Barley to Buy— Of course, any maltster can make sure of his malts by selecting only those barleys which are perfectly matured, and in the case of pale malts this is usually the best policy, because the feeling of security is worth paying a little extra for where so indefinable a characteristic as quality is the chief desideratum. For malts for coloured beers, however, where rather more forcing is allowable, the policy of buying only very mature barleys is apt to be rather a costly one. In fact, the really economical selection of barleys for coloured malts is probably the most difficult of the maltster's tasks. Of course, if he is malting for sale, or for a brewery where extract is considered above all things, he is pretty well tied to mature barleys, but where the quality of the malt is the first consideration, the yield of extract becomes a mere matter of *£ s. d.*, or, more correctly speaking, pence and fractions of pence. With a few notable exceptions, it has become the fashion to consider the extract obtained from 336 lbs. of malt as the standard. This is all very well for brewers, but for maltsters and brewers who make their own malt, the real point which should be considered is the extract obtained, not from the quarter of malt, but from the amount purchased, i.e., in most cases from 448 lbs. of barley.

Maltster's Increase and Decrease— This point was brought up in a paper by Dr. E. R. Moritz and myself on "The Economics of Brewery Malting" (*Journal of the Institute of Brewing*, vol. xi. No. 6, July-Sept. 1905), in which the cost of one pound of brewers' extract, including the price of barley and malting expenses (taken for the purpose at 5*s.* per quarter of malt made) was referred back to the barley. To take two examples from the tables printed with the paper: A barley costing 32*s.* (screened in steep) was found to yield 95 lbs. of extract, and each 100 quarters barley (at 448 lbs.) produced 98 quarters of malt (336 lbs.), the cost of each pound of extract referred back to the cost of barley (and malting expenses at 5*s.*) being 4.77*d.* In the case of a barley of the same price and yielding a similar extract on the malt, but producing 102 quarters malt at 336 lbs. for each 100 quarters barley at 448 lbs., the cost of each pound of extract, again referred back to the barley and the cost of malting at 5*s.*, was 4.58*d.*, a difference of 0.19*d.* per lb. of extract, or at 95 lbs. of extract about 1*s.* 6*d.* per quarter between the two barleys.

In fact, assuming that the quality of the two malts was equal, and the value of the first one quoted was 32*s.*, that of the second was 33*s.* 6*d.* If on the other hand the latter was only worth 32*s.*, the former was bought at 1*s.* 6*d.* above the proper value. It may be said that a difference of 4 percent in yield of malt from barley is exceptional, and perhaps it is in any one season where English barley only is considered and that bought from the same district; but a difference of 2 percent between maltsters is fairly common, and that would mean a difference in value of 9*d.* per quarter on the barley bought. When foreign barleys as well as English are considered, the difference is very much greater, ranging between an increase of about 10 percent on dry mellow foreigns, and a decrease of 4 to 5 percent on really damp English, which have to be kiln-dried and stored before being steeped.

Of course the yield of malt from barley is determined chiefly by the moisture percentage with which the former comes to hand, but the relative maturity of the barley also has a considerable influence on the question, mellow barleys losing less than hard ones. Broadly speaking the malting loss due to matter dissolved out in steep, respiration, and rootlet growth averages somewhere about 11 to 12 percent, calculated on dry matter for the barleys likely to be malted by the ordinary sale maltster and brewer. With English barleys of high moisture percentages the matter is generally complicated by the sweating of the barley, but as the moisture content falls below 16 percent an increase of rather more than 1 percent for each 1 percent of moisture less in the barley may be expected.

With the dry thin foreign barleys increase is always counted upon when purchases are made and the amount of increase may be computed with a very fair prospect of accuracy by the moisture percentage as determined in the laboratory. English barley, however, is usually purchased in comparatively small lots on the open market and the estimation of moisture must be a more empiric matter. In practice it is usually guessed at by the bite of the barley, but it should always be remembered that a ricey or steely barley will seem drier to the bite than a mealy one of equal moisture percentage. Mealy barleys also have a somewhat lower specific gravity than steely ones and are slightly better value for this reason.

Hard Barleys which Improve and Hard Barleys which do not Improve— It has been stated, perhaps with undue reiteration, that some apparently ricey barleys mature during sweating and storage, or sometimes without sweating during storage. If these barleys are soaked in water for six or seven hours, and subsequently slowly dried at low temperatures, the change in their appearance and in the condition of the starch cells of the endosperm is apparent. This artificial maturation has, so far as is known, never been carried out in practice, the operation being somewhat long and costly, but a very similar process probably goes on in wet seasons, when barleys are bought with moisture percentages of 17 to 20 and sweated at low temperatures on the kiln, the barley in these cases almost invariably showing all the signs of complete maturation after sweating unless really steely when bought. When, however, barley is sweated in dry seasons, the moisture present is not sufficient to effect any perceptible amount of maturation. Now a large percentage of the English barley crop comes to hand undermatured in most seasons and whether this barley can be conveniently artificially matured or not depends to a great extent upon the amount of water it contains, so that in a wet season a much larger proportion is likely to make up into tender malt than in a dry one where sweating is not necessary to ensure germination. On the one hand, in a wet season the price of barley will be comparatively low, and if carefully bought and sweated a great deal of it may be made into tender malt, the yield of malt from barley being low. In a dry season on the other hand, the price of barley is likely to be higher, and while the yield of malt from barley will also be comparatively high, the necessity for selecting well-matured lots of barley will be greater if the malting process is to be an easy one, and the malt tender.

Water Bought as Barley— Of course, when English barley comes to hand with a high percentage of moisture, the excess of water bought as barley must be taken into consideration. Thus, taking the average moisture content of English barley at 15 percent and considering a barley with 20 percent of moisture, the deficiency on value owing to high moisture content would amount roughly to 1*s.* 6*d.* per quarter on barley value where this was about 28*s.* per 448 lbs. on the dry sample. On malt value, which in reference to dry matter is obviously higher owing to the lesser weight of the standard malt quarter, the difference would be nearly 2*s.*, and to this would have to be added the expense due to fuel and labour for sweating, and the loss of dry matter which would be likely to accompany the maturing process occurring when sweating a barley of such high moisture content.

On the other hand, the barley would probably be less dear to buy owing to its damp condition, and if skilfully bought would probably after sweating be very well matured material and, as such, likely to yield tender malt and high extract.

Out of Condition and Heated Barleys— In wet seasons a certain amount of barley often comes to hand in bad condition, i.e., either slightly heated or with a tendency to heat which has been arrested—this material should always he rejected, for malting purposes, as even if the vitality is actually unimpaired it is not likely to make up into really sound malt. Actually heated samples, of course, should always be rejected, but their appearance is by no means confined to very wet seasons, when farmers are likely to foresee the danger and take special precautions against its occurrence.

Grown-Corns— Experience is the greatest factor needed for the detection of heated corns in a sample of barley; they are generally characterised by a rather "foxy" colour on the skin above the germ, and the whole sample generally shows signs of the sweat it has gone through by its extremely mature appearance. Apart from samples coming to hand with heated corns and generally out of condition, there are often some grown-corns present in barley when the season has been a wet one or when the barley has been for several months in the rick. If the barley is dressed over by the farmer after thrashing, many of these are often blown out, but even if a small percentage be present, it is not necessarily advisable to reject the barley which contains them if it is to be kiln-dried and made into malt for running beer, as after drying and storage they will probably germinate again and make up into fairly tender malt.

Nitrogen Content— Before referring to the different sorts of barley available for malting and attempting roughly to assess their relative values as malting and brewing material, it may serve a useful purpose briefly to consider the criteria present to aid the purchaser in his selection of raw material. Moisture content has already been considered and apart from this the albuminoid content of barley generally gives some slight indication of its adaptability to the flooring process. Whether the total nitrogen content, or the percentage of soluble nitrogenous matter present is of most importance is perhaps still doubtful, but the broad fact remains that barleys relatively low in total nitrogen content as a rule are better suited for the manufacture of tender malts than those of relatively high nitrogen content, although different varieties vary considerably in this respect.

Appearance— Probably the greatest help of all is the appearance of the endosperm starch, but it must be borne in mind that the origin of the barley is of great importance in this context. For instance, three samples of barley very similar in outward appearance and condition of endosperm starch might be found, one being imported Chevallier Chilian, one imported Chevallier Californian, and one a white dry home-grown barley from Bedford or Huntingdon. On being asked which sample he would select for malting, any one who had had any experience in malting barleys from the different localities named would probably place the Chilian first, the English second, and the Californian third, although their resemblance to each other might well be such that he would be at pains to select the sample he liked best on appearance alone. In other words, soil and climate have a preponderating influence on the malting quality of barley, and the difficulty of the grower lies not only in selecting seed which is in itself of intrinsic value, but also in studying what varieties are best adapted to the particular soil and climate in which they are to be grown.

CHAPTER X

MALTING LOSS

A Certain confusion of ideas has existed, and probably still exists, in many quarters on the subject of malting loss. In attempting to estimate it and compare the loss on different barleys three facts must be borne continually in mind.

1. That a common weight must be assumed for barley and malt, say, by reckoning the quarter of malt at 448 lbs. barley weight, and not 336 lbs. malt weight.
2. That commercially dry malt contains a varying percentage of water, and that any quantity of absolutely dry malt, say 110 quarters will amount, with a moisture content of, say 2 percent, not to 112 but to 112.24 quarters of commercially dry malt.
3. That in order to make any sort of comparison between barleys of varying moisture content, malting loss must be expressed on dry matter and dry matter only.

All this may seem self-evident, but every maltster is not a trained scientist, and it may be in some cases difficult to realise that two barleys which show identical malting loss referred to dry matter as barley, but which vary in moisture content actually lose the same percentage of dry matter during the malting process.

In this chapter the expression raw barley will be used to denote barley either kiln-dried or not, dry barley will mean dry matter in barley, so that 100 quarters of raw barley having a moisture of 15 percent would represent 85 quarters of dry barley.

Malting loss is best expressed by equations, and below are given two equations:

(1) Of a mellow English Chevallier containing 16 percent of moisture.

(2) Of a mellow Brewing Californian containing 10 percent of moisture, and each showing the same loss calculated on dry matter.

	Raw barley		Water		Loss		Dry malt
(1)	100	=	16	+	9.24	+	74.76
(2)	100	=	10	+	9.90	+	80.10

In each of these cases the malting loss calculated on dry matter is exactly 11 percent, but as 100 raw barley represents in the first case 84 dry matter, and in the second case 90 dry matter, though in the equation (which represents in each case loss on raw barley) the loss is at first sight higher for the drier barley, it is apparent on a closer scrutiny that the two losses are identical (9.90 : 90 : : 9.24 : 84).

It is therefore obvious that it is absurd to say that one barley, for instance a dry foreign, shows a higher loss in, say, rootlet growth than another, which may be a comparatively damp English, when 100 of one may represent 89, 90, or 91 dry and 100 of the other 84, 85, or 86 dry.

This point has been emphasised, it is hoped not unduly, because one so often hears remarks such as "foreign makes more rootlet than English," or "more soluble matter is steeped out of one sort than another," simply because there are great differences in the moisture contents of raw barleys.

Setting aside the contained water of barley and the loss of weight due to its expulsion during drying operations, malting loss proceeds from three sources.

1. Soluble matter lost during steeping.
2. Rootlets.
3. Respiration.

Expressed on dry matter the maxima and minima losses are probably about:

	Maximum	Minimum
Respiration	6.7	5.7
Rootlets	4.5	3.7
Steep	1.8	0.6
Total	13.0	10.0

and calculated on a barley with a moisture content of 15 percent, the figures would be

	Maximum	Minimum
Respiration	5.7	4.8
Rootlets	3.8	3.2
Steep	1.5	0.5
Total	11.0	8.5

Loss by Rootlet Growth— This can always be approximately estimated by weighing the malt culms screened from a kiln-load of malt of which the barley has been weighed into steep, but it must be remembered that besides the rootlet screened out, a certain quantity will have been knocked off during kilning and fallen through the apertures in the kiln floor; the amount of the latter will vary considerably; according to the nature of the drying-floor, and according to whether the malt has been trodden or not before being thrown off. Treading is done generally where the malt is screened by hand, and is not usually necessary when a rotary malt screen is used. **Loss by Respiration** need not be discussed here as long as it is remembered that it is chiefly influenced by the quality of the barley and the duration of the flooring period, the amount of sprinkle liquor given and the temperature at which the growing pieces are kept, all factors which have a corresponding effect on rootlet growth.

Loss in Steep— A most interesting paper has recently been written by Professor ADRIAN BROWN (*Journal of the Institute of Brewing*, vol. xiii. No. 7) on the steeping of barley, which sheds new light on the question of soluble matter steeped out of barley.

Professor BROWN's investigations have made it clear that the testa of the barley corn—a very fine membrane surrounding the endosperm and the embryo of the barley corn—is permeable by water but not by many solutions of mineral acids, including a 58 percent solution of sulphuric acid.

This being so it is argued, and subsequently practically proved, that none of the matter contained within the testa can escape through it into the steep, which of course means that all the matter dissolved out of barley in steep must be dissolved from the portion of the corn outside the testa, viz., the pericarp and pales. In fact, the loss in steep consists not of the starch which is potential brewers' extract, but of the husk which is potential grains, so that while the maltster loses weight to the extent approximately of from 0.5 to 1.5 percent, this loss does not directly affect the brewer in the extract yield obtainable from the malt.

The loss in steep therefore probably increases with the huskiness of the barley, and this fact obviously increases the value of thin-skinned as compared with thick-skinned barleys.

In the discussion following the paper referred to above, Professor Brown quoted Escombe's work for the Guinness Research Laboratory, showing that barleys heavily manured with nitrogenous manures are likely, by reason of their robust growth in the field, to form highly developed pales, which have the effect of producing the husky appearance in the thrashed barley commonly referred to as thick skin.

Now barleys containing a high nitrogen content, are, as has been pointed out by many writers, unlikely, other things being equal, to modify so satisfactorily upon the floors as those with low nitrogen content, and we should therefore look to the typical thin-skinned barley of low nitrogen content to lose less weight during the malting process than typically thick-skinned barley of high nitrogen content. Of course when malting loss is being considered, it must be taken for granted that the resulting malt is sufficiently tender for the purpose for which it is made, undermodified malts naturally weighing heavier than tender malts.

It has been suggested above that 13 percent calculated on dry matter is about the maximum loss likely to be incurred on the flooring system for the barleys likely to be malted for the English brewer, and that 10 percent calculated on dry matter is about the minimum, but it is a subject on which it is impossible to state rigid limits, and the above maxima and minima losses are suggested with all reserve.

Below are given some figures actually obtained practice on large bulks of barley weighed into the cistern and off the kiln:

	Raw Barley		Moisture		Dry Malt		Loss	Loss calculated on dry matter
(1)	100	=	9.0	+	81.3	+	9.7	10.6
(2)	100	=	9.5	+	81	+	9.5	10.5
(3)	100	=	12.0	+	78.3	+	9.7	11.0
(4)	100	=	14.0	+	8.9	+	8.9	10.3
(5)	100	=	16.0	+	75.0	+	9.0	10.7

Where

(1)	=	938 qrs	No. 1 Standard Brewing Californian	1905 crop
(2)	=	470 qrs	Brewing Californian	1905 crop
(3)	=	1270 qrs	Chevallier Chilian	1906 crop
(4)	=	210 qrs	English Archer (a very fine barley)	1906 crop
(5)	=	1800 qrs	English (Chevallier and Archer)	1906 crop

Perhaps it need hardly be written that the dry malt figure is multiplied by 4/3 and the moisture added to ascertain what may be called the commercial increase. That is taking (5) and assuming a 2 percent moisture in the malt, 75 x 4/3 = 100 dry malt or 102.04 quarters commercially dry malt were obtained for each 100 of raw barley steeped.

Similarly allowing a malt moisture of 2 percent, the commercial increase on (2) was actually 10.2 percent, of (1) 10.6, of (3) 6.5, and of (4) nearly 4.9. No. 4 was a very fine thin-skinned Archer, and shows an extremely low malting loss, though not so low as some which I have found in Brewing Californian, Brewing Chilian, and Spanish, and it is perhaps open to doubt whether malting loss on two-rowed barleys is on the average quite so low as the loss on six-rowed barleys.

The 1000 Corn Test— It may be as well to give a description of the so-called 1000 corn test which, if carefully carried out, affords a very accurate criterion of malting gain or loss.

The screened barley is very thoroughly and carefully sampled before being steeped; a large representative sample being taken. A moisture determination is made, and not less than 5000 corns counted out, and the average weight of 1000 noted.

When the finished malt has been screened and cooled to the normal temperature, it is similarly sampled, its moisture content determined, and the average weight of 1000 corns noted, not less than 5000 corns being counted and weighed.

Now supposing 1000 corns of barley weigh 42 grams, and 1000 corns of the resulting malt weigh 31.5 grams, we shall have exactly 3/4 barley weight for malt, or each 100 raw barley will yield exactly 100 malt at malt weight, or there will be no gain or loss between the barley steeped and malt produced.

This will be sufficient for the commercial outlook, and will tell the maltster all he wants to know with regard to the cost of his malt, and if two or three tests are made about the beginning of a season of characteristic barleys, the results may be depended on to give a very fair indication of the net gain or loss at the end of the season as long as barleys continue to come in with about the same moisture content. If, however, the test is undertaken with a view to finding out what characteristics in barley tend to produce low malting loss, it will be necessary to carry it rather further.

Supposing in the case we have taken the moisture of the barley as steeped was 16 percent, and that of the malt 1 percent, then

42 grs. barley (moisture 16 percent) = 31·5 grs.
malt (moisture 1 percent); or
35·28 grs. dry barley = 31·185 grs. dry malt; or
100 grs. dry barley = nearly 88·4 grs. dry malt;
malting loss on dry 11·6170

The test itself is rather a tedious one, whether the corns are counted automatically or not, and as has been seen, requires to be very carefully carried out, especially as to sampling, and if it is to be depended on for question of cost it is needless to say that all the barley to be malted must be very similar in general character and moisture content to the lot upon which the test is made.

To sum up, the characteristics likely to result in low malting loss are thin skin, mellowness accompanied as is almost always the case by low nitrogen content, and general fine malting quality.

Size and shape are probably immaterial, except in that in very thin barley the proportion of husk to starch is likely to be high, and malting loss correspondingly heavy, owing to loss in steep; on this ground, large barleys would be preferable to small were it not for the fact that very bold barleys are generally rather husky ones.

With regard to weight, light barleys are obviously likely to lose less than heavy ones and the value attached in some quarters to large, bold heavy barleys is probably quite a false one.

CHAPTER XI

MALTING COST

THERE are three courses open to the brewer for filling his requirements in the way of malt. He may buy it in the open market, he may pay a maltster a commission for buying barley for him and malting it, or he may buy barley and malt it for himself.

The first of these methods needs little comment; it has become fairly usual to buy malt on analysis, and if this is done and deliveries are regularly checked in respect to general quality, cleanliness, moisture content, and diastase, tint and extract, there should be little danger of trouble arising in the brewery caused by malt.

Sampling— When malt is sampled by the brewer it is only fair to the maltster that a really representative sample should be taken, say a portion from every tenth sack in a delivery, and that the sample should be immediately placed in a clean tin or stoppered bottle pending analysis.

The second method, i.e., buying on commission, is not quite such a simple matter, and entails rather more faith on the part of the brewer in the skill and honesty of the maltster. It is usual for the maltster to buy his barley delivered into his maltings, the brewer having previously intimated the price he wishes to pay and the quality he desires. On the delivery of the barley, bulk samples are sometimes sent to the brewer with the price paid and quantity bought, and, if the samples are approved, a cheque is returned, though more often the brewer does not see the barleys bought, but simply intimates the price to be paid and leaves their selection entirely to the maltster. When the barley has been malted it is stored apart and deliveries are sent to the brewery as required, further payment being made for the amount of the commission as may be agreed on. Where English barley has to be sweated a slight extra charge may be made, and arrangments have to be made for the disposal of barley screening and malt culms.

For instance a brewer, A, may arrange with a maltster, B, to buy 1000 quarters of English barley for him at about 30*s.* per quarter, screen it, and make it into malt for him at, say, 5*s.* per quarter on out-turn of malt carriage from malting to brewery being payable by the brewer. Supposing sweating to be unnecessary, and that all the barley is bought on one day and delivered within a week. On approval of the bulk samples B sends A a cheque for £1500. It is agreed that the barley should be malted as soon as possible, say, in a 100-quarter house, and taken in lots of 50 quarters at a time, deliveries to be weekly, and to commence six weeks from the purchase of the barley, commission being payable on completion of contract. Then the transactions will be recorded by accounts something like the following:

Dr. to B & Co., Maltsters

Messrs. A & Co., Brewers.

		£	s.	d.
Oct. 1. 1000 quarters barley at 30*s.*		£1500	0	0
Jan. 8. Commission at 5*s.* on 960 quarters malt		240	0	0
Less 20 quarters screenings at £1	£20 0 0			
10 quarters broken corn at 25*s.*		12	10	0
5 tons malt culms at 70*s.*		17	10	0
		50	0	0
		£190	0	0

It is not to be supposed that 5*s.* is a standard price for commission malting. Possibly 5*s.* 6*d.* is more usual. About 6*d.* is generally allowed for sweating barley, and it is often arranged that the receipts for sale of malt culms and screenings should be payable to the maltster and not to the brewer.

It will be seen that apart altogether from the satisfaction given by the malt, there will be several figures which A will have no opportunity of checking, even without considering those cases complicated by sweating barley and any loss of weight caused thereby, and if A is not satisfied with the malt and would like to refuse it, he is in a very different position (having paid for the barley and tacitly approved of the malting method adopted) from the brewer who has bought his malt on an analytical standard not maintained in deliveries.

Neither can it be argued that the brewer will know exactly what his malt will cost him on the commission system, this depending largely (and especially in the case of dry, thin foreign barleys) on the return of malt from barley, which will depend to a very great extent on screening loss and malting loss.

It is, perhaps, necessary to say at this point that not the slightest shadow of suspicion is intended to be thrown on the many maltsters who make up malt on this system. On the contrary, the very fact that implicit confidence is placed on their skill and judgment is the best proof possible that they fully deserve it, but in considering the expenses of malting it is necessary to refer to the commission system on which many brewers buy, and in so doing it is necessary to point out its possible drawbacks as well as its advantages; there are several of the latter. The brewer who has a good knowledge of barley and can arrange with a maltster whose maltings are situated in a district in which the barley suits him can get his barley bought by the man on the spot, and can practically in most cases have the malting done according to his own ideas. Thus, it is common in many cases that the maltster should undertake not to make up any malt, say, before October 1, or after the end of April, and should furnish charts showing the temperature at which the floors are worked, and how much water was given in steep and sprinkle. The brewer also has the advantage of the good buying of the maltster. Thus, supposing a maltster working on commission finds he can buy the wished-for quality of barley below the limit price, the brewer has the advantage over him who bought his malt outright, in which case the fall in price would benefit the maltster, although as a rule it is probable that the latter will prefer to pay the full price and buy a better article rather than pay less and take the risk of turning out malt of which his client may not approve.

In the case instanced above all the barley was taken as having been bought at the same time. Of course this is a matter for arrangement between the contracting parties; the barley purchases may be spread over four or five months as arranged, but it will be obvious that in any case the maltster will be paid for his barley almost as soon as he pays for it, so that no interest on capital expended on barley will be borne by him. Apart from that, the payment for deliveries of malt may be weekly, monthly, or as arranged, and of course each contract may be differently arranged in respect of sale of screenings, malt culms, &c. The commission system varies in different cases, and it is only possible to give a general idea of it, but it is probably chiefly carried on in small and old-fashioned maltings in country districts where rent and rates are low and the brewery supplied situated fairly near the maltings.

No doubt it sounds as though the maltster who is paid five or six shillings per quarter on malt made can make very little profit on the transaction, but it must be remembered that he has to pay practically no interest on money expended on barley, and buys his barley delivered in to the maltings, that he has no carriage to pay on malt, and that as a rule the rent of the houses in which malt is made on commission is very low.

The brewer who makes his own malt generally does so for one of three reasons: either the brewery has maltings attached where it has always been the custom to make up malt, or he thinks he may economise by building or leasing maltings and saving the maltster's profit, or he considers that by having direct control over purchases of barley and the system of manufacture he will obtain a malt better suited to his individual requirements than could be purchased on the market.

At the time of writing, the malting trade is passing through a period of depression, and the price of malt has been cut down in many cases to a figure, which, far from showing profit, makes it only just possible for the maltster to pay interest on his capital. With raw material varying so infinitely as barley, it must be obvious that the maltster's first concern must be to buy his barley as cheaply as possible, and it is only natural that cut prices in malt should be followed by cut prices in barley, with the result that in very many cases the quality of the malt made is as low as is consistent with analytical or other requirements.

In this state of affairs the price obtainable for malt must necessarily depend to a great extent on the reputation of the maltster, and there are many names, which, in themselves guarantees of quality, command full and even extravagant prices from those brewers who make it a rule that the quality of their materials should be first class. On the other hand, there are maltsters ready to supply first-rate malt at low prices—prices, in fact, quite inconsistent with a fair return on capital either because their capital is sunk in maltings, or because they expect matters to mend and are ready to lump their loss for a time, being able to afford to wait for their profit until they have established a reputation, the trade has improved, or their less wealthy competitors have been forced to close down.

Yet, again, there are the less scrupulous gentry who are content to sell on a fine sample at a cutting price and deliver malt which will bring in a good profit, trusting to the negligence or cupidity of the brewers to pass inferior deliveries.

So, after all, the position of the brewer in the market for malt is not such a very easy one. He may buy on a reputation, in which case he will probably have to pay for it. Or he may buy as cheaply as he can, in which case he may be fortunate in getting more than value for his money, or unfortunate in getting extremely uneven deliveries.

Basing the cost of malt on the price at which malt can be bought it is improbable that any brewer can make as cheaply as he could buy, but where really consistent deliveries are required it is probable that substantial economies can be effected if care and knowledge are expended on the purchase or erection of maltings, or where suitable existing buildings can be leased and the purchase of barley and the process of manufacture carefully controlled.

There is undoubtedly a danger where quality is the first desideratum that over anxiety should cause the buyer to be somewhat extravagant in his purchases. Enough has been written about barley to make it clear that its purchase is not an easy matter where both economy and quality have to be studied, though it is very easy to ensure either at the expense of the other, and it is therefore very tempting for the buyer to make certain of his quality, arguing that it is better to pay a shilling or even eighteen-pence more in order to be assured of a really fine malt than to risk turning out a mediocre malt for the sake of a comparatively small saving in cost. Of course this is true as far as it goes, but the judgment of the buyer must come in when determining exactly to what extent caution may be wisely carried, and it is probable that the profits of many brewery maltings suffer from over caution on the part of the buyer. Of course recklessly low buying is much more objectionable, for if a brewer turns out malt that he cannot use himself he is not likely to find a purchaser for it at a price above that of pig-food.

Again, it is surprising how many brewers who make their own malt are utterly ignorant of the cost of malting. Two shillings and sixpence, three shillings, and three shillings and sixpence added to the cost of barley seem to be favourite answers from these gentlemen, who pay for their barley out of brewery capital and seem to think the only expenses of malting barley are labour, fuel, and water. On asking a most successful brewer how much he charged per quarter of the malt he made for rent, the reply was that he would ask the auditors, and on these gentlemen being consulted it transpired that the maltings having been built out of brewery profits it would not be fair to make any extra charge. In fact, as far as profit and loss was concerned, the capital expended in building had been simply written off.

A head brewer, in fact, is generally a very busy man, who in many cases has little respect for what he sometimes considers the quibbles of the accountant, while the latter has not always all the practical knowledge necessary for the adjustment of accounts, so that when the brewer buys the barley and generally superintends malting operations it is perhaps not unnatural that the actual cost should not be quite clear.

As a matter of fact the cost of malting is not a very simple matter. It is complicated first by the question of out-turn. Thus, if the cost of making 100 quarters of barley into malt is £25, the cost per quarter of malt will obviously depend largely on the number of quarters of malt obtained, which may vary between, say, 97and 107.

From this it appears that all calculations should be based not on the barley bought, but on the malt obtained, taking the malt quarter at 336 lbs., and not the barley quarter at 448 lbs., as the standard unit.

The full charges which must be set against the number of quarters of malt turned out at the end of the season's operations are as follows:

Rent payed on maltings or, say, 5 percent on expended capital.

Rates and Taxes.

Depreciation where the buildings are owned at, say, 2 percent on outlay and where machinery is used, a higher rate, say, 7 to 10 percent on capital expended on it.

Before dealing with the other expenses it will be as well to remember that the money set aside under the foregoing heads may be regarded as a standing charge, and will be independent of the amount of malt turned out.

To make this clearer we will take the case of a house designed for an output of 12,000 quarters of malt per season and built at a cost of £24,000, of which, say, £2000 is expended on machinery. The standing charges will be:

Rent, say, 5 percent on £24,000	£1200
Depreciation, say, 2 percent on £24,000	480
Rates and taxes, say	220
Depreciation on machinery, say	200
	£2100

or, divided by 12,000 quarters of malt made, 3*s*. 6*d*. per quarter. Now, supposing that from one cause or another only 6000 quarters of malt were made in any season, the cost per quarter on standing charges would at once mount to 7*s*. per quarter. Of course this is an extreme case, but it is intended to bring out the extreme necessity of economy when maltings are built.

To continue the list of malting expenses, and presuming that the business is to be financed as most malting businesses are, on an overdraft from a bank, that barley is paid for on delivery, and three months' credit given for payments on malt, then *interest on working capital*, viz., barley purchases, wages, fuel, and sundry expenses will generally cost about 1*s.* per quarter.

Fuel may vary from 9*d.* to 1*s.* 3*d.* according to current prices, kiln construction, and the skill and attention of the stoker.

Where machinery is used power may cost from 1*d.* to 3*d.* per quarter, the former figure being rather a low one, water from 3/4*d.* to 1*d.*, and general expenses, including insurances, &c., somewhere about 2*d.*

Labour, of course, must vary immensely, the actual labour for malting alone, apart from supervision of men and machinery, varying roughly between 9*d.* and 1s. 3d. per quarter.

Interest on floating capital.

Labour.

Fuel.

Power, water and general expenses, then, should cost somewhere between three and four shillings per quarter of malt made, so that the total cost on the 12,000 quarters in the house referred to would probably be somewhere between 6*s.* 6*d.* and 7*s.* 6*d.* per quarter of malt made, from which would have to be deducted somewhere about 6*d.* per quarter from the sale of malt culms and kiln dust, leaving a net cost varying between 6*s.* and 7*s.* per quarter of malt produced.

In order to arrive at the cost price of the malt, this figure must be added to the cost in barley value of each quarter of malt made, that is, the total amount expended on barley must be divided into the number of quarters of malt produced from that barley, the latter figure being influenced by (1) the amount of screenings taken out less return from screenings sold; (2) the amount of malt made from each quarter of barley steeped.

Screening Loss— Supposing 102½ quarters of barley be bought at 30*s.* per 448 lbs. delivered into the malt-house, the cost of screening should be reckoned in the following manner:

102⅝ quarters at 30*s.* cost		3075*s.*
1 quarter broken corn brings	20*s.*	
1 quarter thin corn brings	15*s.*	
4/8 quarter dust and dirt brings	---	
		35*s.*
100 quart clean barley cost ready to steep		3040*s.*

Cost per quarter ready to steep, 30.40*s.*, or nearly 30*s* 6*d.*

Now if these 100 quarters of barley produce exactly 100 quarters of malt, the cost price of the malt will be between 36*s*. 6*d*. and 37*s*. 6*d*.

This case was made as simple as possible, but we will now consider the case of:

A. A very damp English barley, which had to be sweated before steeping.

B. A very dirty, dry, thin foreign, such as, say, a Syrian Tripoli

A.—1. Sweating.

100 quarters at 28*s*. cost		2800*s*
Sweating expenses at 6*d*. per quarter		50*s*.
Sweating loss 10 quarters		---
90 quarters dried barley cost		2850*s*.

Cost of dried barley, 31*s*.; 8*d*. per quarter.

2. Screening.

90 quarters at 31s. 8d. cost		2850*s*.
1 quarter broken at 20s.	20*s*.	
1 quarter thin at 15*s*	15*s*.	
0.5 quarters dust		
		35*s*.
87.5 quarters screened barley cost		2815s.

Cost of barley screened and dried 32*s*. 2*d*. per quarter.

3. Malting

87·5 quarters screened barley cost	2815*s*.
92 quarters malt produced at (say) 6s. per quarter	552*s*.
92 quarters malt cost	3367s.

Cost per quarter, 36*s*. 7*d*.

B.—1. Screening.

100 quarters of barley at 25*s*. delivered cost		2500*s*.
5 quarters thin at 10*s*. bring	50*s*.	
1 quarter broken at 15*s*. brings	15*s*.	
2 quarters stones and dirst	---	
		65*s*.
92 quarters screened barley cost		2435s.

1 quarter screened barley costs 26*s*. 6*d*. nearly.

2. Malting.

92 quarters screened barley cost	2435*s*.
100 quarters malt produced at 6*s*.	600*s*.
100 quarters malt cost	3035s.

Cost per quarter, say, 30*s*. 4*d*.

Both these are typical cases. In the first there is a heavy loss from barley to malt, though owing to sweating this is not very apparent in actual malting operations, the malt obtained actually showing a considerable increase on the barley steeped.

In the second case there is a heavy screening loss, balanced by a heavy gain from barley to malt, and in both cases careful analysis of screening records, and in the case of the English, of sweating records, is necessary.

Of course, when clean foreign is bought, the gain is obvious, but screening loss is none the less rather a difficult matter to deal with in malting accounts. On a house malting all foreign, the deficit very easily amounts to an average of 5 percent and arises the price of the barley by 4*d.* to 8*d.* per quarter, and in these cases screening sales should be considered as an item in the barley account, the quantity of screenings sold and cash obtained for them being deducted respectively from the totals of barley bought and cash paid.

Thus the last page of the barley ledger would read somewhat as follows:

Date	Sellers	Folio	Quality	Quantity	Price	£	*s.*	*d.*
	Forward		No. 1 F.	9000		12,150	0	0
Apr. 2	Brown & Co.	Lr. 3	do.	1000	27/6	1,375	0	0
		Sales bk.		10,000		13,525	0	0
Deduct	*Screenings*	5		400	12/-	240	0	0
	Waste and Loss			100				
				9,500		13,285	0	0

Cost of screened barley, £17*s.* 11½*d.*

Where, however, as is usually the case, English and foreign barleys are malted in the same house, it is not always possible or even desirable to keep the screenings separate. In most seasons English screenings are a very small quantity, the broken corn separated by the half-corn cylinders being the chief loss, and so it is often as well not to attempt to average out the price of screened barley, but simply to add an approximate figure, say, 3*d.* or 4*d.* for screening loss, and to lump returns from sales with those from sales of malt culms, kiln dust, and any other by-products sold.

The amount of malt culms produced per quarter of malt varies to a certain extent with the barley malted and the flooring method employed, but most maltsters now aim at growing a good thick bushy root, and probably it will be near enough for all practical purposes to reckon from 12 to 13 lbs. of clean culms per quarter of malt made. In addition to this there will be some 3 lbs. of the so-called kiln dust—i.e., culms which have got separated from the malt in the kiln and fallen through the interstices of the drying-floor on to the floor of the hot air chamber. These latter weigh rather heavier than the clean culms subsequently separated by the malt screen, as they are generally mixed with a certain amount of coal dirt and ash, blown up from the fire. They are generally considerably scorched and, therefore, dark in colour. Clean malt culms vary in value according to their cleanliness and colour (pale culms generally making rather more than dark ones), between 70*s*. and 90*s*. per ton, kiln dust averaging between 35s. and 45s. per ton, and taken together they bring in a return of between 5*d*. and 6*d*. on the quarter of malt made.

Thin dry foreign barleys probably give rather more culms per quarter of malt than English, and though the difference is small the question of quantity of culms seems great in relation to the barley steeped owing to the greater amount of malt turned out from equal quantities of English and dry foreign barleys.

It will be seen that the cost of making one quarter of malt may vary somewhere between 6*s*. and 7*s*. in fully equipped modern maltings where the malt is charged with all liabilities and depreciations. Where the out-turn is very large expenses can be cut down to a certain extent, but it is most improbable that malt can be made in any modern house for 5*s*. per quarter *where fair interest is charged on capital expended and further charges made for depreciation*. Malt can be made for 5*s*. per quarter and less in many parts of the country where old houses are rented at low rentals and the maltster's liabilities end with keeping the house in fair repair.

In such cases where 1*s*. or 1*s*. 6*d*. covers rent and repairs, fuel may cost 1*s*. 6*d*., labour 1*s*. 6*d*., and interest on floating capital 1*s*., making a total of 5*s*. to 5*s*. 6*d*. Deducting receipts for sales of culms and assuming that there was no machinery and water was pumped from a well on the premises, the total cost would vary somewhere between 4*s*. 6*d*. and 5*s*. per quarter of malt made. When we consider that labour is likely to be cheap in country districts, rates low, and barley close at hand, it is strange that the small houses are not more in use, yet all over the country we find them unused, superseded by larger and more expensive buildings, and the small maltster driven out of the trade by the competition of his neighbour who works on a larger scale.

CHAPTER XII

BARLEYS

Introductory. *The Large-Berried, Two-Rowed Varieties*— In order to recognise the essential characteristics of the different varieties of barley used in this country for malting it will be necessary briefly to examine the ear of barley and note its varying structure.

The main criteria of maltability in barleys are maturation and low nitrogen content, thus, for instance, if two varieties of English barley be taken, say Kinvers and Halletts, and it be assumed that one variety is superior to the other as malting material, granted both have been successfully grown and matured, and then a mature sample of the inferior barley be compared with an immature sample of the superior barley, the malting value of the two may be reversed.

Beaven puts this so concisely in his "Varieties of Barley" (*Journal of the Institute of Brewing*, vol. viii. No. 5, July and August 1902), that it may be permitted to quote him at length.

"When barley values are assessed the effect due to good maturation and commonly called kindliness or mellowness is nearly always associated with the plump round contour of the transverse section which the eye instinctively recognises. *But no one variety or sort of barley* (the italics are mine) is always good or is always better than another even in this respect, and it is most necessary to avoid generalisations. Comparisons must always be qualified with the proviso all other conditions being equal, and in actual practice this almost never happens."

Apart from this, however, certain varieties are undoubtedly better suited to certain climates and soils than others in the average of seasons, and the maltster who can differentiate between such differ-

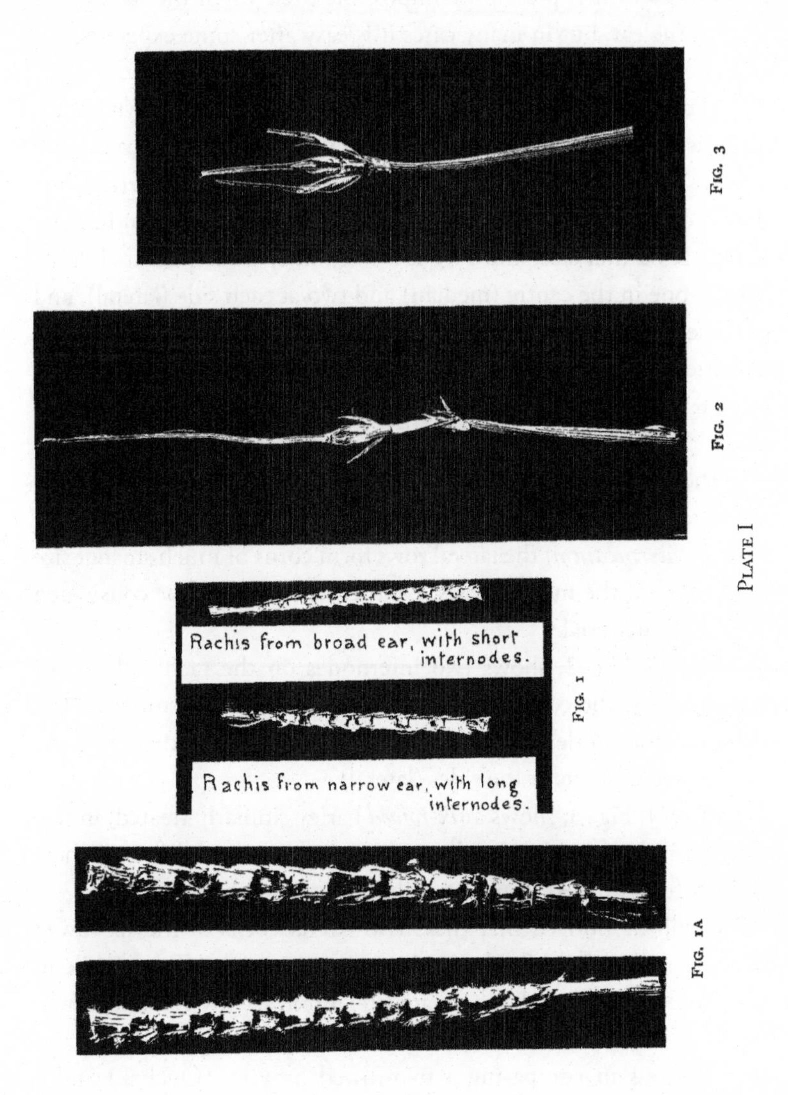

FIG. 3

FIG. 2

FIG. 1

FIG. 1A

PLATE I

ent varieties certainly has an advantage in the long run. This is not always possible on the market when the threshed barley only is seen. In some cases it is practically impossible even when the barleys are seen in the ear, but in many cases it is easy after some experience to recognise different varieties as seen in sample on the market.

The part of the stem from which the corns grow is known botanically as the *rachis.* This rachis is divided into divisions of varying size known as internodes placed on opposite sides of a theoretical line drawn down its side. From each of these internodes (shown in Plate I, Fig. 1, natural size, and Fig. 1A magnified) three potential corns spring, one in the centre (median) and two at each side (lateral), and of these three corns one (the median) may be fertile and form a grain of barley, and two (the lateral) infertile, in which case the barley is known as two-rowed, or all the grains may be fertile, and form grains of barley, in which case the barley is known as six-rowed.

There are also two varieties, in one of which (*decipiens*) the lateral rows are rudimentary and without floral organs, and in the others of which (*intermedium*) the lateral rows form corns of much smaller size than those of the median rows. Neither of these need be considered as malting material.

Plate I, Fig. 2, shows two internodes on the rachis of a two-rowed barley, the corn having been stripped from the top one. On the lower internode is seen the fertile in the centre (median) and the two unfertile florets at the sides (lateral).

Plate· I, Fig. 3, shows a *six-rowed* barley similarly treated, in this case it will be seen that in place of two infertile lateral florets, there are two fully formed lateral corns besides the median one.

It will be noticed that these two lateral corns of the six-rowed barley are curved, while the median corn is straight. Thus, while in two-rowed barleys all the corns are straight, in six-rowed barleys two-thirds of the corns are curved and one-third straight. This fact is very obvious in comparing a two-rowed Smyrna (Ouchak) and a six-rowed Smyrna (Yerli) (*see* Plate VII).

The Two-Rowed Barleys— The barleys grown in England and used for malting consist entirely of the two-rowed variety. Six-rowed barleys are grown in the United Kingdom, but up to the present time

Chevallier

Archer

Plate II

these have been used entirely for forage or grinding, although it is just possible that in the future as increased attention is paid to the race-improvement of the cereals, six-rowed varieties of barley may be grown in this country of quality suitable for malting.

If ears of two-rowed barleys of different varieties be closely examined, it will be seen that, whereas in some cases the internodes on the rachis are short and the grains stand out from the ear, in others, the internodes are longer and the corns lie in close to the ear.

These varying structures of the rachis are shown in Plate I, Fig. I, natural size, and Fig. 1A enlarged. They indicate appreciable differences of variety, and it has become common to allude to the former as broad-eared barleys (*Hordeum Zeoc riton* var. *Erectum*) and the latter as narrow-eared barleys (*Hordeum distichum*).The latter are generally loosely classed as "Chevalliers." Strictly they may be divided into—

1. True Chevalliers—Kinvers Halletts, &c. (Plate 11A).
2. Archers (Plate 11 B)
3. .Hannas.

The latter are for some reason not much grown in England, and have never become favourites, but they are much grown in Hungary and Moravia, and some extremely fine samples of them are sometimes imported from those countries.

At this point it becomes necessary to turn for a moment to the botanical side and further examine the grain. The portion of the corn lying next to the stem is known as the ventral side and the portion lying away from the stem as the dorsal side. Running down the length of the ventral side is a furrow, known as the ventral furrow, and lying in the furrow, attached to the lower (embryo) end of the corn, is a rudimentary secondary axis (*rachilla*), known commonly as the basal bristle; different forms of the basal bristle are fairly constantly associated with different varieties of barley, and although a magnification of three or four diameters may be necessary to make the differences clear, it is easy for most men whose eyesight is good to perceive the main difference between different varieties in this way. The basal bristle may be long with short hairs (*villous*), or practically hairless (*glabrous*), or short with long hairs (*hispid*). The latter variety is associated with both Archers and Hannas, and the second (long hairless)

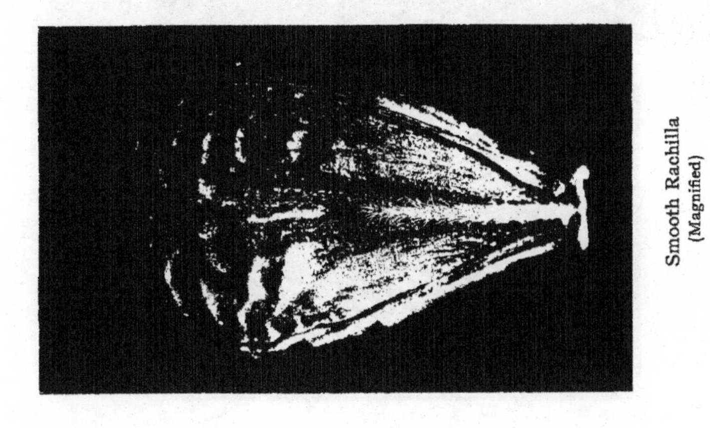

Smooth Rachilla
(Magnified)

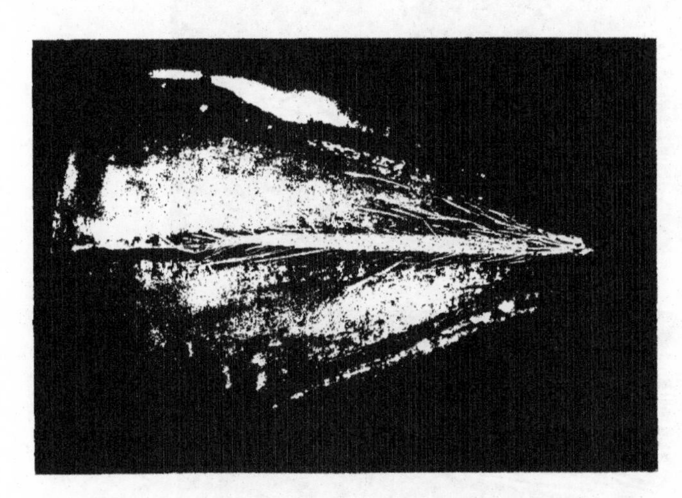

Hairy Rachilla
(Magnified)

PLATE III

Fan Barley
(Zeocrithum)

Spratt

Goldthorpe

Plate IV

variety with Chevalliers (Kinvets, Halletts, &c.).

Now both Archers and Chevalliers (Kinvers, Halletts, &c.) are widely grown in England. With regard to the latter some soils are better suited to one variety and some to another, but it is doubtful whether the thrashed barley could be distinguished as belonging to any one of the sorts of Chevallier distinctively named by seedsmen as Kinver, &c., and they will therefore be grouped as Chevalliers in this chapter.

It is possible, however, to distinguish between Chevalliers and Archers both by the basal bristles, and after considerable practice by the appearance, and, maturation being equal, I am inclined to prefer Archers as malting material. This however, is not probably the general opinion, and Chevalliers are certainly rather bolder as a rule.

Of the broad-eared British barleys two varieties stand out. These are Goldthorpes, under many different names, such as Invincible, Burton Malting, and Standwell (the latter originally grown from a cross between Goldthorpe and Chevallier), and Spratts.

Both varieties may be easily distinguished from each other and from the narrow-eared varieties, and both generally have hairy basal bristles, although a new sort of Goldthorpe has recently been introduced which has a smooth basal bristle. Goldthorpes are grown chiefly in Yorkshire and further north; they are as a rule larger in the berry than either Chevalliers, Archers, or Spratts, and their skins have a peculiarly shiny appearance. After a certain amount of experience they can readily be distinguished from the other varieties.

Spratts are also early distinguishable from the other types; they are grown chiefly in Essex, Hertfordshire, and the Lincolnshire fens, and the grains are very short and round with hairy basal bristles and a well-defined, almost circular enlargement of the ventral furrow towards the top of the grain (*see* Plate V); when well matured they make excellent malting material, though they are exceptionally difficult to malt when immature. Both Goldthorpes and Spratts probably appear somewhat coarser in the skin than the narrow-eared barleys, though, on the other hand, Goldthorpes are advantageously grown on soils which would not produce satisfactory barleys of the Chevallier and Archer types.

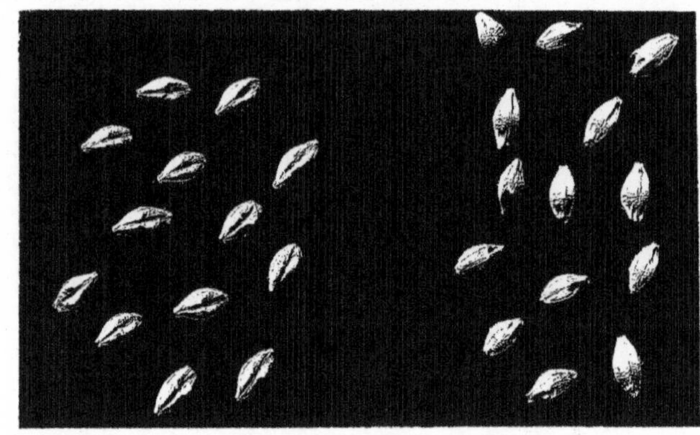

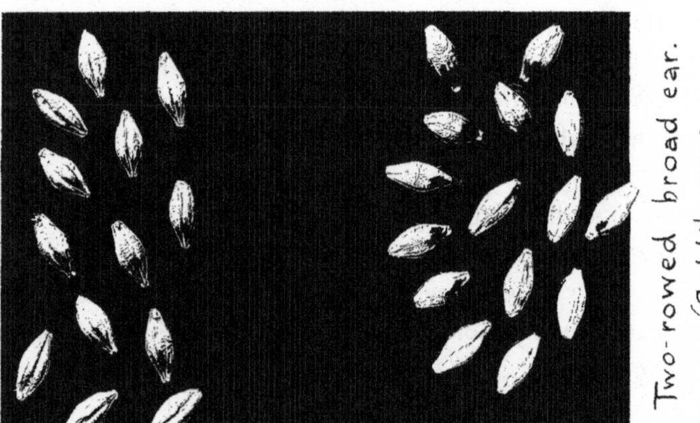

PLATE V

The terms thick-skinned and thin-skinned have been freely used in the text as they are expressive and generally accepted. In reality the pales do not vary in diameter, at any rate for barleys of the same country, although loosely fitting pales give the appearance of thick skin, and *vice versa.*

With regard to the quality of the malt turned out, and assuming the greatest perfection of which each variety is capable, the order would probably be: 1, Archers; 2, Chevalliers; 3, Goldthorpes; 4, Spratts; although, judging solely by lowness of nitrogen content, Goldthorpes might very possibly come first.

As regards extract yield per quarter of dry barley the order would probably be : I, Goldthorpes; 2, Chevalliers, 3, Archers; 4, Spratts; extract yield, other things being equal, being obviously a matter of the size of the berry.

The finest barleys as regards quality are generally to be found on light loamy and drift soils overlying either (a) the chalk formations and upper greensand, (b) the ower greensand, which often outcrops on the edge of the chalk, (c) certain of the oolitic formations, which may be broadly grouped as gravels.

If a geological map of England be examined it will be seen that there are two narrowish strips of these formations running roughly in the form of a curve from north-east to south-west. The "gravels" strip is the largest and lies to the north-west of the chalk strip, and between the two there is often a very narrow strip of the lower greensand.

The gravels beginning some few miles south of the mouth of the Tees in Yorkshire run due south with only one interruption past Brigg over Lincoln Heath, Sleaford, Grantham, Stamford Heath to Bedford. From there they branch south-west and continue in a belt some twenty miles broad past Aylesbury, Oxford and Chippenham on the south, and Buckingham, Woodstock, and Minchinhampton on the north, narrowing to the width of a few miles at Shepton Mallet and branching out again further south, finally to finish in Chesil Bank and Portland Bill.

The chalk formation starts between Hunmanby and Flamboro' Head in Yorkshire, runs south-west and then due south through that

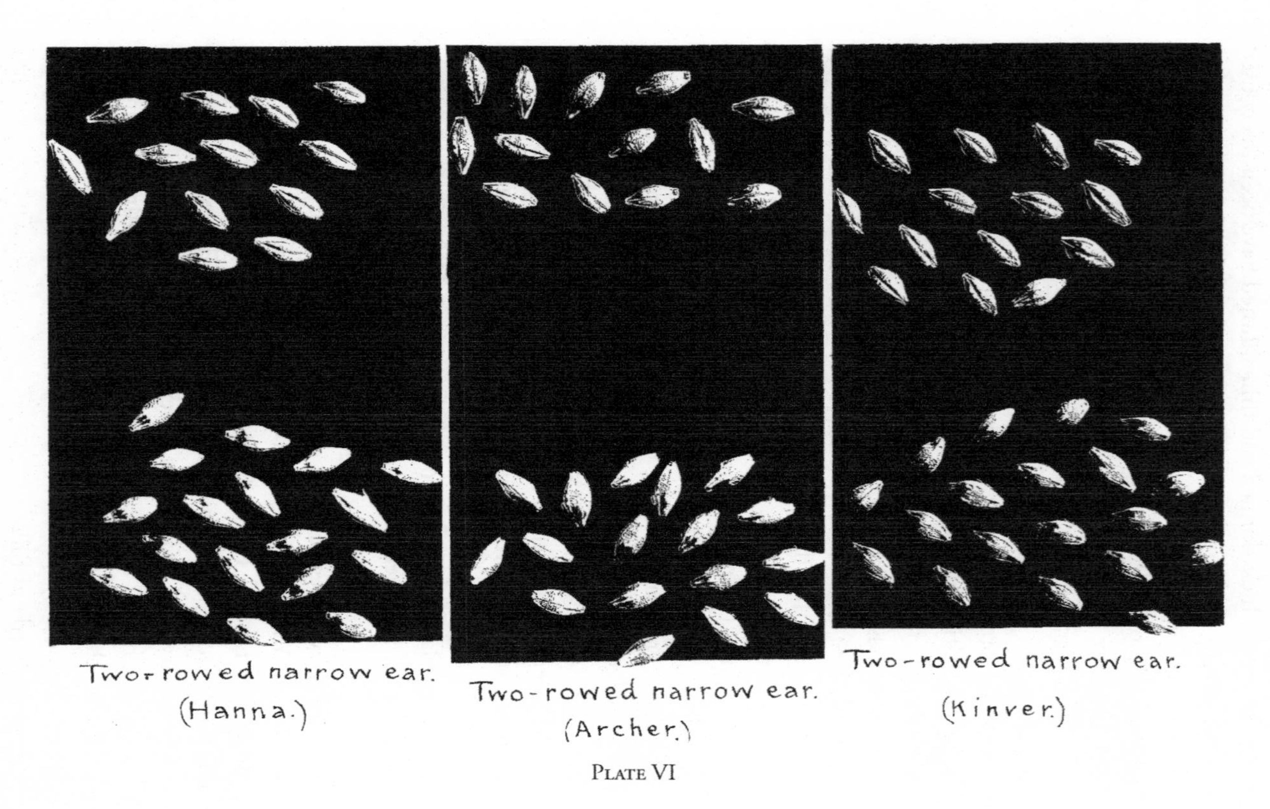

Two-rowed narrow ear. (Hanna.)

Two-rowed narrow ear. (Archer.)

Two-rowed narrow ear. (Kinver.)

Plate VI

county to the north coast of the Humber, continues through the wolds of Lincolnshire where there is a somewhat broad outcrop of lower greensand from Caistor to Burgh, and recommences in Norfolk about Hunstanton, from which point it trends due south on a strip about twelve miles broad to Newmarket whence it branches southwest past Cambridge, Royston, Baldock, Berkhampstead and Reading to Marlborough, at which point it broadens out to the Salisbury Plain and Shaftesbury down country in the west and the down country of Hampshire to the east. East and west of Amesbury there are two strips, the westerly one running down to the narrow strip of gravel country bordering Chesil Bank, and the easterly one running due east on a broad strip of some twenty miles to Farnham on the north, and Petersfield on the south, from which points start the North Downs and South Downs, the latter stretching to Beachy Head and the former through Surrey and Kent to Dover.

Of course it is not to be imagined that these strips include all the barley land of England. They undoubtedly include most of the best. The county of Durham grows excellent barley of the Goldthorpe type in many seasons. The east coast of Norfolk and a great part of Suffolk are famous for their barleys, and Essex which is shown on the geological map as "London clay" almost invariably grows very useful barley; in fact all the East Anglian counties are largely and often deeply overlaid with drift.

If it be permitted to generalise, the finest narrow-eared barleys are grown on the drift and light loamy soils over the lower greensand, upon similar soils over chalk, and next over gravel. The broad-eared barleys are probably not so dependent upon soil, and although the finest of them are probably found on the gravel and chalk-covering soils of Yorkshire, those grown upon heavier soils as a rule produce barleys of much greater malting value than could be expected from narrow-eared barleys grown under equally adverse conditions.

In addition to the two-rowed large-berried barleys grown in the United Kingdom, there are several countries which export barleys similar to our own in fair quantities in the average of seasons. These are France, Germany, Hungary, the U.S.A. and Chili.

The finest barleys imported from these countries are probably seldom, if ever, superior as raw malting material to the finest home-grown barleys, but apart from the economic reasons which may occasionally cause any country to import barleys which can be bought at prices comparing favourably from the buyers' standpoint with home-grown material, there are climatic considerations which may sometimes cause the British brewer to consider the advisability of using a certain amount of imported barley.

There are, for instance, a good many brewers who think it necessary that the barley malted, especially for pale beers, should show "sun," that is, that it should have been subjected during its last month or six weeks in the field, and especially during harvest, to hot sunshine. It would be extremely rash in the present state of our knowledge on the subject to criticise this opinion. It may be correct; on the other hand, many maltsters consider that barleys harvested after a wet English summer and carefully selected and kiln dried form malting material equal to anything obtainable. Certainly if it is necessary to buy barleys one year and malt them the next, it is desirable that dry barleys should be selected. Little is definitely known as to the effects of storage on barley, but no doubt certain hard barleys mellow very considerably during a year in store; while many kinds of barleys, bought damp, kiln dried and stored show decreased vitality after prolonged storage periods.

In discussing the imported two-rowed large-berried barleys little attention will be paid to their variety, as their malting quality is primarily affected not by variety but by the conditions of soil, climate and manuring under which they are grown

France— The barleys imported from France are generally rather smaller than the average of English, the finest (Sablais) are in favourable seasons very kindly, but on the whole, French barleys are rather hard. The extract obtainable from them is considerably lower than that obtainable from English.

Germany and Hungary— The finest barleys imported from these countries are of the Hanna type, and in favourable seasons are equal in quality and extract yield to the finest English Chevalliers and Archers obtainable after a good harvest. The second qualities are generally inferior to second-rate English barley.

Two-rowed narrow ear.
(Ouchak.)

Six-rowed narrow ear."
(Yerli Smyrna.)

PLATE VII

Chili—Chevallier Chilians are generally of extremely fine quality and naturally very considerably lower in moisture content than English, the barley malts easily and makes malt of very fine quality, comparing favourably in extract yield with the finest English Chevalliers.

California— Chevallier Californians are very similar in appearance to Chevallier Chilians, both barleys being generally bleached almost white by the sun. The resemblance ends there, however, as Chevallier Californians are ordinarily extremely difficult to modify on the floors, although if they can be successfully malted the quality of the malt made is very good. When well modified the extracts obtainable compare fairly well with high-class English Chevalliers, but the malt very constantly is hard-ended and consequently low in extract.

The moisture content of barleys varies of course with the harvest. In dry years such as the harvest of 1906 some English barley came to hand with moisture content of 12 to 13 percent. The average may be taken as about 16 percent French, German, and Hungarian are on the average slightly lower. Chevallier Chiliart and Chevallier Californian both come to hand in this country with an average moisture content of about 12 percent, and during a year's storage they pick up about 1 percent of moisture.

The Imported "Brewing" Barleys— As the two-rowed barleys are divided into two main varieties, consisting of broad-eared and narrow-eared, so the six rowed barleys may be subdivided into broad-eared (*Hordeum Hexastichum*) and narrow-eared (*Hordeum Vulgare*).

By far the greatest proportion of imported barley of the so-called "brewing" type belongs to the latter, while a very much smaller proportion is of the former variety. Yet, again, a certain amount of two-rowed narrow-eared barley is imported which is classed as "brewing" barley, and a considerable quantity of barley is imported and sometimes used as brewing material which is obviously a mixture of two-rowed and six-rowed.

It has been pointed out in a preceding chapter that it is easy to distinguish between threshed barley of the two-rowed and six-rowed varieties, and if reference is made to the photographs of Ouchak and

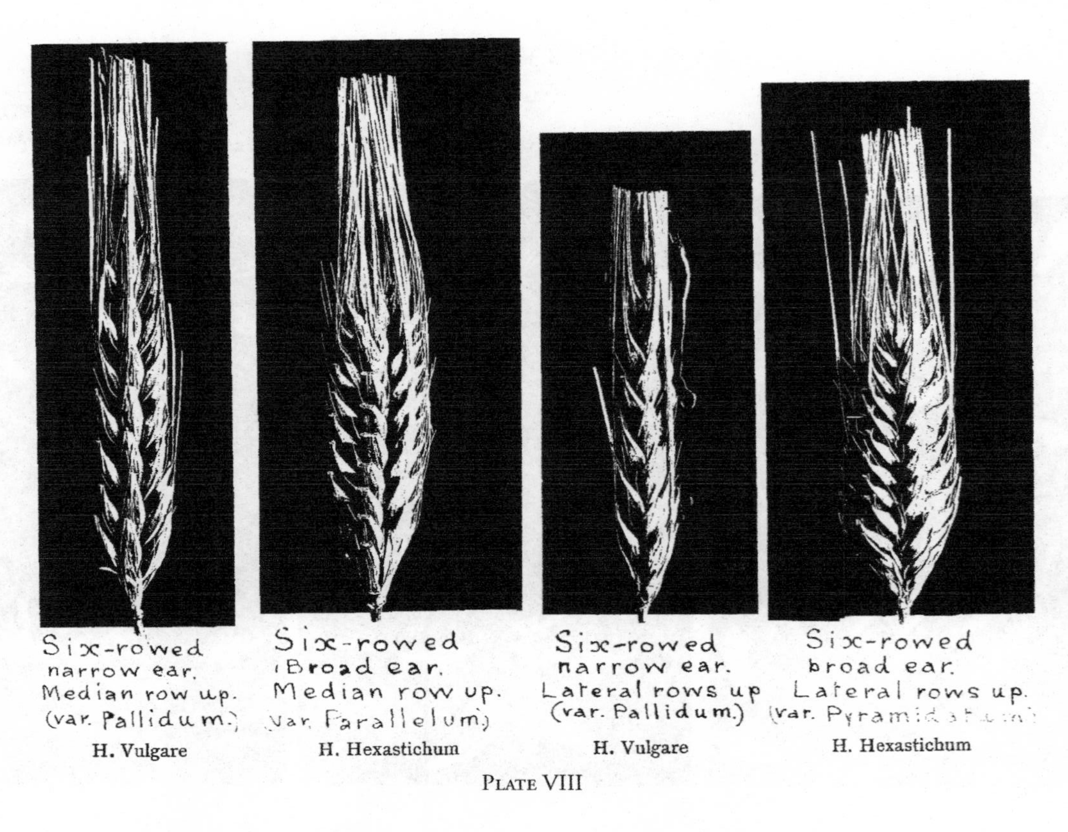

H. Vulgare H. Hexastichum H. Vulgare H. Hexastichum

PLATE VIII

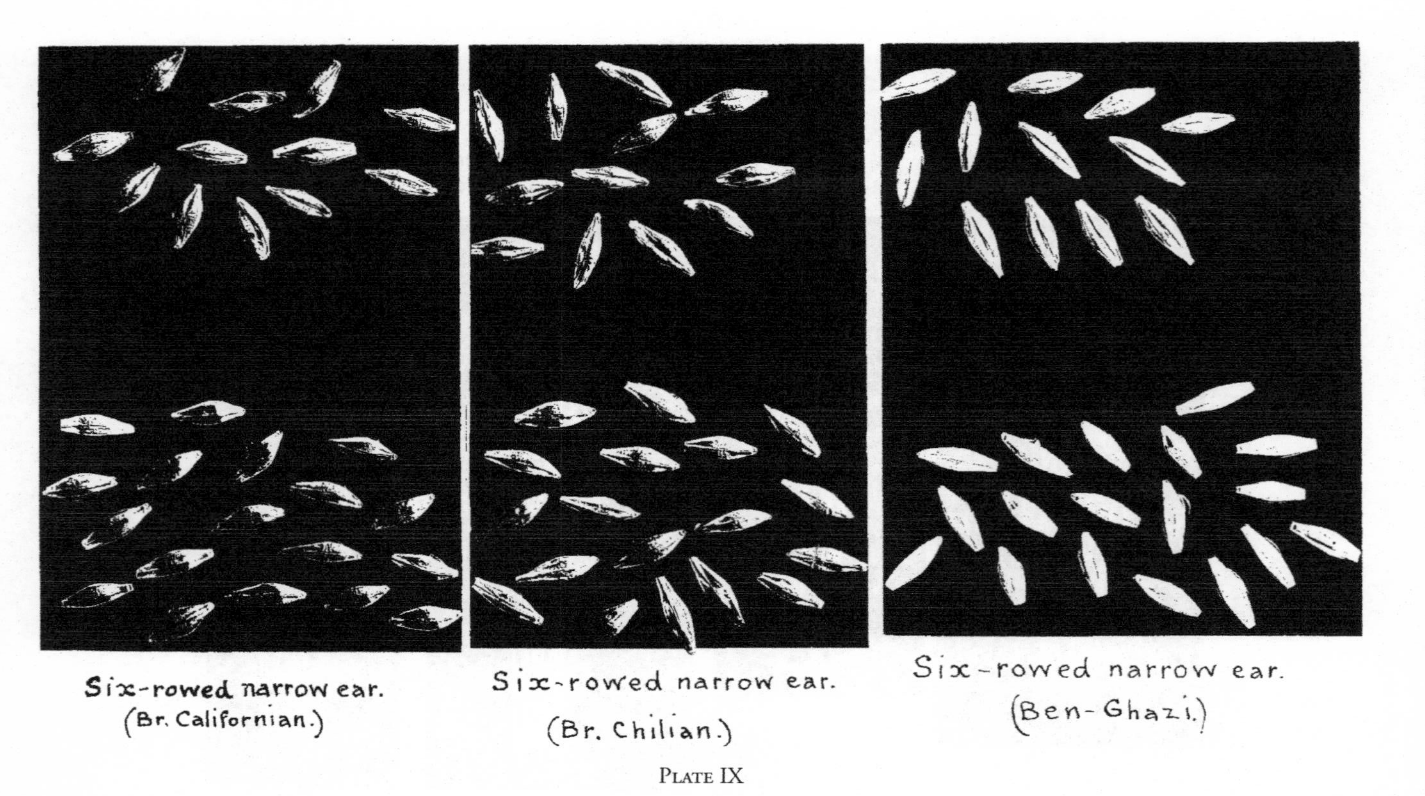
Six-rowed narrow ear.
(Br. Californian.)
Six-rowed narrow ear.
(Br. Chilian.)
Six-rowed narrow ear.
(Ben-Ghazi.)

Plate IX

Yerli Smyrna it will be seen that the twisted corns of the latter, which amount to 66 percent of the whole, are easily seen; but there is no such easy means of distinguishing between threshed barleys of the broad and narrow-eared six-rowed type, both of which have the same proportion of twisted corns, and which vary in size enormously. Ben Ghazi barley, for instance, which commonly belongs to the narrow-eared type, is characteristically much smaller than brewing Chilian which is frequently of the broad-eared variety.

Still larger is Tunisian, a narrow-eared barley, and yet larger the French and Belgian Escourgeon, broad eared.

Again, no criterion is afforded by the basal bristle, which varies greatly, yet the races are quite distinct, as may be seen by a comparison of the photographs (*see* Plate VIII).

In considering the imported barleys commonly alluded to as "brewing barleys" it will be best to divide them into three varieties, viz.:

Hordeum distichum (two-rowed).

Hordeum Vulgare (six-rowed narrow ear).

Hordeum Hexastichum (six-rowed broad ear).

Two-Rowed— The imported narrow-eared two-rowed barleys consist of Ouchak, sometimes divided into Anatolian and Marmora Ouchaks, but all of the same main characteristics, and Syrian Tripoli, often known as Hama.

The latter are characteristically long thin barleys, though the size varies, and yellowish in colour. They come to hand very dry, 11 to 12 percent of moisture, and, as a rule, containing a high percentage of thin corn and red sand and dust. The latter of course weighs heavily, and being worthless detracts a good deal from the value of the barley. The thin corn should preferably be screened out and malted separately, when it may form useful brewing material. Hama barley, as a rule, malts well and yields very fairly tender malt. The extract obtainable must, of course, vary with the size and quality of the barley and the success of the malting method adopted. The malt when ground in the Seck mill at the 25 setting should yield from 85 to 88 lbs.

In giving approximate figures for the extract yielded by foreign brewing barleys the Seck mill at the 25 setting will be taken as the

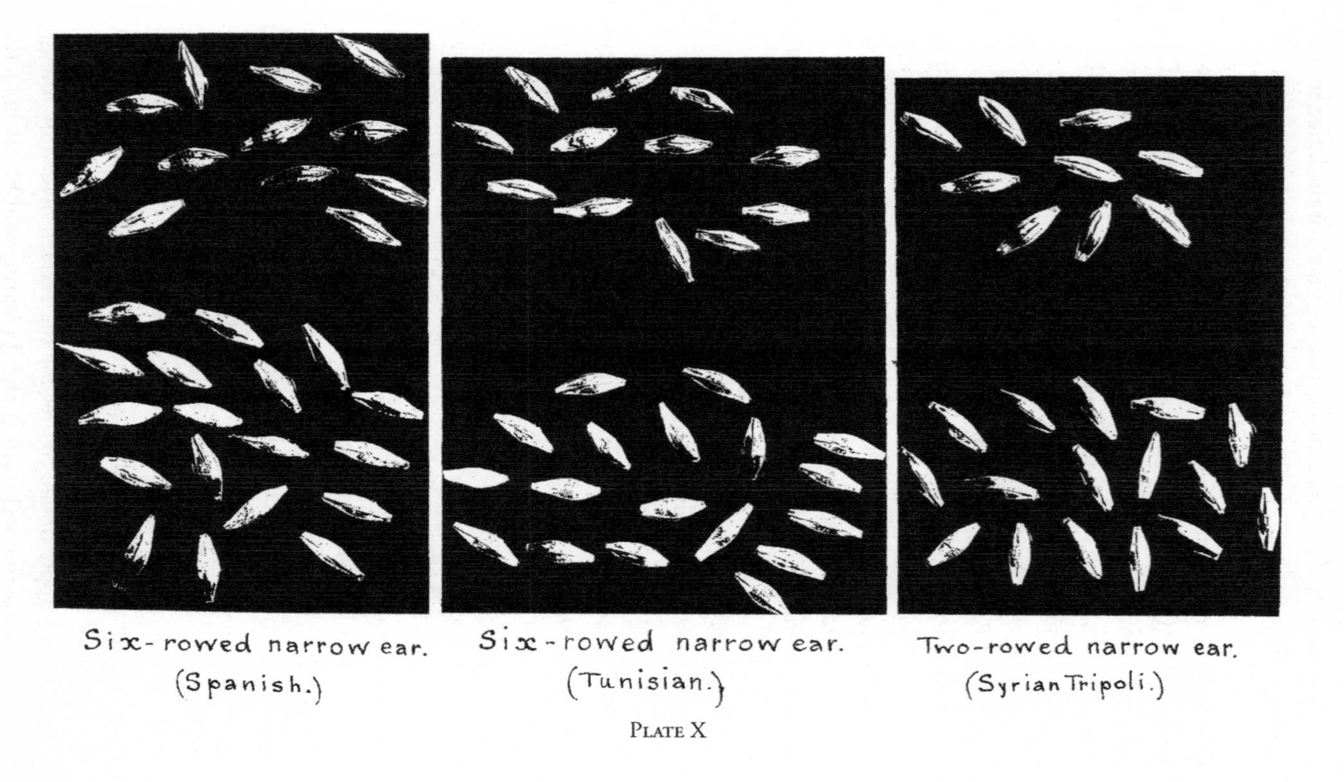

Six-rowed narrow ear. (Spanish.)

Six-rowed narrow ear. (Tunisian.)

Two-rowed narrow ear. (Syrian Tripoli.)

PLATE X

standard grind.

Ouchak barleys vary enormously in both size and appearance. Some come to hand no larger than Hama, some nearly as large as English. In colour some are almost dead white, some are mixtures of black and white corns, and some are yellow. The moisture content varies approximately between 11 and 13. Their cleanliness varies, as some are screened prior to export.

The barley, however, is characteristically steely and perhaps the most difficult of any barley to make into tender malt. The vitality of the barley is, as a rule extremely strong, and it can consequently be steeped for very long periods, in some cases 100 hours and over with advantage, but the malt is rarely tender and extracts consequently rule low considering the size of the corns, 90 lbs. being seldom attained even at the fine grind which of course favours steely malts.

The imported barleys of the six-rowed narrow-eared type come from many countries and vary very much in malting quality, although as will be seen from the accompanying photographs the appearance of barleys from different countries is often very similar.

It will probably be best to give a list of them and subsequently consider each on its merits.

They include:

Brewing Californian	Tripoli Coast (African)
Oregon	Algerian
Brewing Chilian	Morocco
Spanish	Gaza
Yerli Smyrna	Cyprus
Ben Ghazi	Azoff
Tunisian	Bessarabian, &c

Brewing Californians hardly need description. They come to hand as a rule very fairly clean with moisture varying between 10 and 12 percent, are easily malted and make very tender malt of excellent palate and yield extracts varying between 88 and 92 lbs.

Oregons are almost identical both in appearance and quality, but on the average of seasons perhaps slightly inferior.

Brewing Chilians as a rule are somewhat larger and coarser in the skin, with perhaps a slightly higher moisture percentage. They do not modify on the floors quite so readily as the Brewing Californian barleys and yield on the average a pound or two less extract. The malt, however, generally forms very fine brewing material.

Spanish barley is not very largely imported into this country and varies in character between Brewing Californian and Brewing Chilian. The extract obtainable from it varies according to quality somewhere between 86 and 90 lbs., but the malt is generally of fine quality.

Yerli Smyrna is usually very largely imported. It varies greatly in colour and moisture percentage with the season. The finest lots are sometimes almost dead white and sometimes yellow, and generally yield malts of fine quality with extracts varying between 86 and 90 lbs. Inferior lots are seldom attractive either in appearance or malting qualities, and this is the fact to such a marked extent with the barley that the best policy is usually either to buy the finest quality or to avoid it altogether. In comparing Yerli with Brewing Californian, it must be remembered that while the former gets to this country often within six or seven weeks from harvest the latter arrives after a voyage of four months or over. Certainly Yerlis usually improve with storage and, sometimes when they come to hand yellow and comparatively damp, kiln-drying may be beneficial.

The North African barleys vary greatly with the season, but Ben Ghazi is usually of very fine malting quality. As against this, however, it is characteristically so thin that the extract obtainable from it is usually very low, 82 to 85 lbs., so that unless it can be bought very well it is not very good value from the economic standpoint. Next in order in point of quality come *Tunisian* and *Tripoli Coast* barley. Both are often extremely fine in appearance and all that can be said about them is that the malt made from them is usually somewhat disappointing, the average extract being 85 to 87 lbs.

Algerian and Morocco are characteristically a little larger, a little coarser, rather less satisfactory to malt, and productive of malts of slightly lower quality and about the same extract value.

Gazas are not usually of very good malting quality. They are also typically very thin and consequently yield low extracts. They may be classed as generally inferior to Algerian and Morocco.

Cyprus exports barley of fine quality and barley of poorer quality. The former does not come on the open market and I have no experience of it except that it looks extremely good. The poorer qualities are not very attractive.

The Black Sea Barleys, Bessarabian, Theodosian, &c., need not be discussed. They are not very attractive in appearance as a rule and seldom produce satisfactory malt

The six-rowed broad-eared barleys are, as has been said, not much imported for malting

Persian is largely imported for grinding, and in some seasons comes to hand of sufficient quality for malting. It is rather thin black and white barley, and generally contains a good deal of thin corn and black oats. It is especially necessary to ascertain that the germination is satisfactory before the barley is purchased, as large percentages of dead corn are often present. As has been said, it sometimes arrives of very fair quality and is then easily malted and makes up into useful second quality malt, yielding in some cases extracts of 87 to 88 lbs. Imported Persian barley, however, is generally partly at any rate, of the six-rowed narrow eared variety.

Brewing Chilian of the broad-eared variety is said to be imported sometimes. I cannot say that I have recognised it, but I have had experience of very large coarse-skinned barley from Valparaiso almost comparable to a French Escourgeon, which was probably of the broad-eared variety. It malted extremely well and yielded an extract of well over 90 lbs.

Mixtures— The barleys imported as grinding material from Russia and the Danube are often mixtures of narrow-eared six-rowed and two-rowed barleys. They are characterised in the main by their very high nitrogen content and seldom make satisfactory brewing material, even when they are pure-bred six-rowed barleys. They are probably better this year (harvest 1907) than ever before.

To summarise, imported barleys of the so-called brewing type may be placed in the following order for (1) quality of malt; (2) quantity of extract, it being understood that the order must necessarily vary in both cases with the different seasons.

1. Quality	2 Extract-Yield
Brewing Californian	Brewing Californian
Ben Ghazi	Oregon
Oregon	Brewing Chilian
Brewing Chilian	Spanish
Spanish	Finest Yerli
Finest Yerli Smyrna	Ouchak
Syrian Tripoli (Hama)	Syrian Tripoli
Tripoli Coast	Finest Persian
Tunisian	Tunisian
Finest Ouchak	Tripoli Coast
Finest Persian	Algerian
Algerian	Morocco
Morocco	Black Sea
Gaza	Ben Ghazi
Black Sea	Gaza
Danubian	Russian
Rusian	Danubian

The popularity of brewing foreign malts is probably due in the main to consideration of cost. Some brewers undoubtedly consider that the grist of thin husky foreign malts promotes drainage in the mash tun and consequently saves extract, but with the perfection now attained in most breweries in grinding, mashing, and sparging operations, it is not probable that the advantage amounts to very much. Again, after wet English harvests many brewers rightly or wrongly desire the presence of a proportion at any rate of sun-dried barley malt in their grists, but the main reason for the use of a high percentage of foreign malt is probably one of £ *s. d.*

Let it be remembered that with almost all the foreign brewing barleys the maltster can count on an increase varying between 6 and 9 percent, as against a maximum increase of about 3 percent and a possible loss of 4 percent on English barley, that foreign barleys can be satisfactorily malted in warmer weather than English and that until the present season (1907-08) the price of foreign barley has been on the average well below that of English barley likely to be malted for similar purposes. The difference in the cost price of English and foreign malts is in fact considerable and allows for a good many pounds difference in yield of brewers' extract.

To consider the case of malts suitable for the manufacture of pale beers and brewing beers made from English and foreign respectively:

1.—	(A) A fine English barley costs per quarter	32*s.*
	Malting at 5*s.* per quarter	5*s.*
	Increase	---
	Cost price per quarter	37*s.*
	i.e., say, 96 lbs. extract cost 37*s.*, or 4·61*d.* per lb.	
	(B) A fine brewing foreign costs per quarter	28*s.*
	Malting at 5*s.* per quarter	5*s.*
	Increase—*deduct*	2*s.*
	Cost per quarter	31*s.*
	i.e., say, 90 lbs. extract cost 31*s.*, or 4.13*d.* per lb.	
2.—	(A) A medium English barley costs per quarter	28*s.*
	Malting at 5*s.* per quarter	5*s.*
	Increase	---
	Cost per quarter	33*s.*
	i.e., say, 93 lbs. extract cost 33*s.*, or 4·25*d.* per lb.	
	(B) A medium brewing foreign costs per quarter	26s.
	Malting at 5s. per quarter	5s.
	Increase—*deduct*	2s.
		29*s.*
	i.e., say, 87 lbs. extract cost 29s., or 4d. per lb.	

It may be that the quality of the malt made from the English barleys is superior to that made from foreign brewing barleys, and in the main this is probably so. Allowing for husk, I have certainly tasted many foreign malts such as Brewing Californians, Oregons, Brewing Chilians, and Spanish, which in tenderness and palate compare well with high-class English malts. Ben Ghazi is so thin that the husk seriously interferes with the palate and in the case of second-grade foreigners, English probably has a decided advantage in the average of seasons.

But even so, it will be seen from the foregoing comparisons that up to the present year, even comparing second quality English with first quality foreign, the latter have held an economic advantage, and it is more than probable that if the cost of the pound of brewers' extract obtained from the brewing foreign malts were to exceed that of the pound of extract obtained from English malts, a much smaller proportion of foreign barley would be brewed in this country.

WORKS REFERRED TO IN TEXT

BAKER, JULIAN, AND DICK, W. D.
"Some Observations on the Steeping of Malting Barley."
Journ. Fed. Inst. Brewing., vol. xi. No. 5. 1905.

BEAVEN, E. S.
"Fuel-consumption in Malt-kilns." *J.F.I.B* , vol, x. No. 5, 1904.
"Varieties of Barley." *J.F.I.B.*, vol. viii. No. 5. 1902.

BROWN, ADRIAN.
"On some Questions concerning the Steeping of Barley." *J.F.I.B.*, vol. iii; No. 7. I907.

HAYNES, .THOS., JUN.
"The Flooring and Kilning of. Malt." (Review Press, Ltd.)

MORITZ, DR. E. R.
"The Valuation of Malt." (Review Press, Ltd.)

MORITZ, E. R., AND HUGH LANCASTER.
"Economics of Brewery Malting.' *J.F.I.B.*, vol. xi. No. 6. 1905.

POSKIN, MONSIEUR.
"Weevil." *Petit Journal du Brasseur.* 1903. P. 1029.
(*Brewing Trade Review*, Jan and Feb. 1905.)

SALAMON, A. GORDON.
"Some Experiments in Malt-making." *J.F.I.B.*,
vol. viii. No. 1. 1902.

WILSON, ROBERT.
"Boiler and Factory Chimneys." (Crosby Lockwood.)

Index

A

Aeration in Steep, 38, 54
- for barley, 29
- on floors, 69

Albuminoids, soluble, uncoagulable of malt, 65, 83
Algerian barley, 116, 166, 167
Anemometer, 96
Angle at which barley runs, 36
Anthracite, 110, 111
Appearance of barley, 125
- malt, 120

Archers barley, 150
Areas for barley stores, 21, 22
- floors, 4
- kilns, 5
- malt stores, 16
- steeping cisterns, 54

Arsenic, 111
Artificial maturation, 49, 50
Asperguillus niger, 81
Automatic weighers, 11, 25
Azoff barley, 165

B

Barley
- effect of barley on resulting malt, 114
- heat in, 53, 124
- moisture, 45, 127
- out of condition, 124
- relative value of, 114, 121
- sweating drums, 46
- sweating kilns, 45
- weeviled, 30

Barley Bins, 21
Barley, types
- Algerian, 166
- Anatolian, 73, 163
- Archer, 150
- Ben Ghazi, 108, 169
- Black Sea, 167
- brewing sorts, 160
- broad-eared, 150, 160
- Californian, 33, 59, 71, 165
- Chevallier, 150
- Chilian, 160, 165
- Cyprus, 167
- Danubian, 167
- Escourgeons, 116
- French, 158
- Gaza, 166
- Goldthorpe, 153, 155
- Halletts, 146, 150
- Hama, 163
- Hanna, 115, 150, 158
- Hungarian, 157, 158
- Kinver, 146, 150
- Morocco, 166, 167
- narrow-eared, 150, 157, 160
- Oregon, 59, 165, 169
- Persian, 167
- Russian, 167
- six-rowed, 148, 160, 163, 165, 167
- Spanish, 59, 108, 131, 166
- Spratt, 153, 155
- Standwell, 153
- Tripoli, 30, 59, 119, 142, 163, 166
- Tunisian, 59, 116, 163, 166
- two-rowed, 148
- Two-Rowed, 146, 148
- Yerli Smyrna, 59, 163, 166

Baskets, fire, 7, 112
Bisulphide of carbon, 30
Bite of malt, 118
Blinds, 62
Boiling in steep, 58
Brewer's extract, 116

C

C02, 71
Californian barley, 160
Caramelisation, 106, 114
Carbohydrates, ready-formed soluble of malt, 65
Carbon, bisulphides of, 30

Cellar floors, 1, 62
Cellulose, 64
Chalk lands, 157
Chevallier barley, 150
Chilian barley, 160
Cisterns steeping, 39
Coke
gas, 111
oven, 112
Colour of barley, 120
culms, 106, 144
malt, 104, 105
windows, 61
Combustion products, 91
Commission malting, 134
Couching, 68
Cowls, 88
Crushing and grinding, 117
Cubic feet per quarter barley, 20
malt, 16
Culms, quantity per quarter, 144
price of, 144
Curing, 104
heats, 85, 108, 110
Cyclone, dust collector, 18
Cylinder
barley-screened, 36, 37
half-corn separators, 37
malt screen, 17
Cytase, 64

D

Danubian barley, 167
Density of air, 97
Deterioration of barley in store, 27
Dew-point, 96
Diastase, 107
Distichum barley, 150
Double floor kilns, 102
Draining, 68
overnight, 59
Draught, 86
down, 89, 98
effects of on permissible drying heats, 94
Drying, 85, 93
barley, 45
drums, 46
kilns, 45
malt, 85
Dry matter of barley, 128
Dust chambers, 18
destruction of, 17

E

Early-harvested barley, 50
Economy in malting, 3
Elevators
barley, 33, 34
green corn, 43
malt, 12
English and foreign barley values, 168
Enzymes, 64
Escourgeon barley, 116
Evaporation on floors, 76
effect on kiln temperatures, 94
Expense of making malt, 134
of sweating barley, 48

F

Ferments
soluble, 65
Fire baskets, 112
night man, 43, 47
Flavour of malt, 107
Flooring, 64
long period, 81
short period, 78
temperature, 66
temperatures, 68–73
Floors
kiln, 89
working, 60
Forcing, 83
French barley, 158
Fuels, 111
consumption, 110
Furnaces, 111, 112
Furrowing, 100
ploughs, 67

G

Garners
steeping, 24

Gas coke, 112
  engines, 33
  plants, 33
Gaza barley, 166
Glabrous, rachilla, 150
Gravel, soils, 155
Greensand soils, 155
Grinding and crushing malt, 117
Grown corns, 125

H

Half-Corns, 81
  corn, cylinders, 37
Halletts barley, 146, 150
Hannas barley, 115, 150, 158
Hard Barley, 123
  ended malt, 118
Heated barley, 124
Heats, floor, 66, 68–72
  barley drying, 45
  kiln, 86, 110
Heavy barleys, 133
Hispid rachilla, 150
Hoppers, 10, 14
  bottomed bins, 16
  cisterns, 39
Hordeum distichum, 150, 163
  hexastichum, 160
  vulgare, 160
  zeocriton, 150
Hungarian barley, 158
Hygrometers, 96

I

Initial cost of kilns and drums for
    barley drying, 48–49
Intake of barley, 23
Interest on capital, 3, 137, 140

K

Kiln
  barley, 47
  construction, 85
  cowls, 88
  draughts, 99, 100, 102
  malt, 6–11, 88
  shutters, 89
  storage under, 9
  turners, automatic, 43, 101
Kilns, 7, 85
Kinver's barley, 146, 150

L

Labour, cost of, 141
Lime, chloride of, danger of, 58
  water for steep, 58
Loading, 99
"Lumping up". *See* rounding up

M

Machinery, 32
  conveying, 27, 35
  depreciation on, 140
Magpies, 114
Malt
  moisture in, 95, 131
  pale ale, 121
  running beer, 115
  storage of, 12
Malt, barley and,, 114
  bins, 15
  extract of, 116
  green, 68, 86
Malting loss, 127
Maltings, site of, 1
Maltsters increase and decrease, 131
Moisture in barley, 45, 127, 160
  green corn, 93
  malt, 95, 131
Morocco, 168
Morocco barley, 166
Moth, Mediterranean grain, 30
Mould, 81

N

Night Men, 43, 47
Nitrogen in barley, 125, 130
Nitrogenous manure, 130

O

Oregon barley, 165, 168
Ouchak barley, 168
Oven-coke, 112
Oxidase, 71

P

Palate of malt, 115
Pale-ale barleys and malts, 114, 120
Pencillium glaucum, 81
Persian barley, 167, 168
Power in maltings, 33
  shovel, 39
Proteolytic enzyme, 64

Q

Quarter of barley, space required for, 20
  malt, space required for, 16

R

Rachilla, 150
Rake, or pull-plough, 67
Ready-formed soluble carbohydrates of malt, 65
Reek on kilns, 103
Rent of maltings, 3, 140, 144
Resistance, 88, 90, 97
Rougher-out, 23
Rounding up on kiln, 109
"Rushing" in young pieces, 70
Russian barley, 168

S

Sampling malt, 134
Saturation of air on kilns, 95
Screens
  barley, 37, 60
  cylinder, 17, 36
  malt, 17
  shaking, 37
Short, flooring period, 78
  steeping, 56, 59
Shutters, kiln top, 89
Sinker-test, 119
Six-rowed barleys, 148
Skin, "thick" and "thin" of barley, 155
  damaged, 81
Sludgers, 18
Smyrna barleys, 108, 166
Spanish barley, 131, 166
Specific gravity of malt, 119
Spratt barley, 153, 155
Spraying, 76
Sprinkling, 72, 75, 76
Steep, aeration in, 38, 54
  duration, 59
  lime water in, 58
  temperature of, 56
  water, hard and soft, 56
Stick-plough, 67
Stoking, 112
Storage of barley, 20
  of malt, 12
Sweating barley, 45
  expenses of, 48
Sweat in young pieces, 69
  in bin, 52
"Swimming in", 60

T

Temperature
  barley drying, 45, 46
  curing, 85, 109, 110
  floor, 66, 68–71, 72
  malt drying, 85
  steeping, 56
  storage, 28
Testa of barley, 129
Test, sinker, 119
  1000 corn, 132
Thermometers, kiln, 93–94
Tint of malt, 105
Tops, kiln, 88
  storage for barley, 21
Tripoli barley, 142, 163, 166, 168
Tunisian barley, 59, 116, 165, 168
Turners, kiln, 101
Turning, barley, 29
  forks, 67
  on floors, 69
  on kilns, 105
  shovels, 67

U

Uralite, 91

V

Value of, barley, 114
  culms, 144

Value of, extract, 51, 116
kiln dust, 144
moisture bought as barley, 124
Velocity, 86, 97
Villous rachilla, 150
Vitrification, 102

W

Waste of heat, 96
Water, absorbed in steeping, 68, 129
in barley, 124
in malt, 95
removed during kilning, 95
sprinkling, 72
steeping, 56
Wax on barley, 58
Weevil, 18, 27, 29
Windows, 61
Withering, 68

Y

Young pieces, 69

Z

Zeocriton, Hordeum, 150

CPSIA information can be obtained at www.ICGtesting.com
Printed in the USA
BVOW08*1038290415

397937BV00007B/49/P